大连海洋大学与上海世语翻译有限公司校企合作研发

海洋文献翻译教程

Translation for Marine Literature

主　编　郭艳玲　曹　斌
副主编　王　倩　李明秋
　　　　陈　烽　林雅琴

大连海事大学出版社
DALIAN MARITIME UNIVERSITY PRESS

图书在版编目(CIP)数据

海洋文献翻译教程／郭艳玲，曹斌主编. —大连：
大连海事大学出版社，2020.8
ISBN 978-7-5632-4023-4

Ⅰ．①海…　Ⅱ．①郭…②曹…　Ⅲ．①海洋学—英语
—翻译—高等学校—教材　Ⅳ．①P7

中国版本图书馆 CIP 数据核字(2020)第 163571 号

大连海事大学出版社出版

地址:大连市凌海路1号　邮编:116026　电话:0411-84728394　传真:0411-84727996

http://press.dlmu.edu.cn　E-mail:dmupress@dlmu.edu.cn

大连永发彩色广告印刷有限公司印装　　　大连海事大学出版社发行
2020 年 8 月第 1 版　　　　　　　　　　2020 年 8 月第 1 次印刷
幅面尺寸:184 mm×260 mm　　　印张:21　　　字数:408 千

出版人:余锡荣

责任编辑:席香吉　　　　　　　　　　　　责任校对:高　颖
封面设计:张爱妮　　　　　　　　　　　　版式设计:解瑶瑶

ISBN 978-7-5632-4023-4　　　定价:59.00 元

前　　言

　　《海洋文献翻译教程》以英语为载体,以海洋文献为依托,使学生在获取海洋领域相关知识的基础上,提高海洋文献的的翻译能力。本教程能拓宽学生海洋科学视野和思路,扩大学生海洋领域英语词汇及海洋文献文体特征等系统知识量,提高学生综合应用海洋英语的能力,满足我国日益增长的海洋事业发展的需要。

　　本教程包括海洋生物工程、渔业资源与环境、海洋产品与加工、海洋机械工程、海洋土木工程、海洋交通工程、海洋信息工程、海洋经济、海洋法律、海洋文化等10章。每章3个单元,每个单元分别包括一篇课文及词汇、短语、专业术语、专有名词、注释、练习。

　　本教程主要特点:(1)着眼海洋,内容丰富。课文均选编于近年来英语国家海洋领域文献资料,知识内容兼具实效性、普适性和专业性。(2)主题鲜明,实用性强。课文用词严谨,语言规范,突出海洋英语的语言特点,追求地道的海洋英语表达。(3)内容依托,特色突出。本教程以特色的海洋文献为依托,有利于激发学生英语学习的积极性,使学生在学习文章内容的过程中习得语言。(4)侧重翻译。本教程的翻译内容涉及十大领域的海洋文献,内容丰富,实用性强。

　　《海洋文献翻译教程》有利于提高学生英语语言应用能力,尤其是翻译能力,使学生在了解海洋领域知识的过程中习得语言。本教程由大连海洋大学与上海世语翻译有限公司校企合作研发,是我国高校教育工作者内容依托教学的适用教材,是本科生及研究生必要的参考书,也可作为企业培训适用教材。本教程对培养适合21世纪需要的兼顾海洋知识和英语应用能力的复合应用型人才具有重要意义。

编　者
2020年8月

Contents

Chapter 1

Marine Biological Engineering

Unit 1
Marine Biotechnology

What to Do in Marine Biotechnology?

1. Marine biotechnology, which represents a small segment of the biotechnology industry in the US, approximately 85 companies or about 7 percent of all biotechnology companies has applications in medicine, agriculture, materials science, natural products chemistry, and bioremediation. It may be defined as the search for commercial uses of marine biology, biochemistry, and biophysics, and it is a **fledgling** field of study having substantial potential. At the simplest level, there is a sense that organisms living in a saline medium, often at high pressures or temperatures, contain biochemical agents that may be of use to industry in marine biotechnology. Most of the world's tropical nations are very well suited to pursue marine biotechnology because of their **proximity** to the oceans and because of their suitable climates. Worldwide, marine aquaculture produced 14 million metric tons of fish in 1991 with a market value of approximately $28 billion. And there is evidence that the European Union is accelerating its investment in marine biotechnology. A research collaboration between the United States and Mexico could yield considerable benefits for both countries, because the United States is experiencing a **boom** in biotechnology while some of the most promising locations in which to perform marine biotechnology field research are in Mexico.

2. **Aquaculture** is the branch of marine biotechnology that is most closely related to agriculture. Demand for seafood worldwide is expected to increase by 70 percent in the next 35 years. Thus, world aquaculture will need to increase production seven-fold by the year 2025 in order to meet the demand. Unfortunately, this increase in demand comes at a time when the world fisheries are over-exploited and/or "commercially extinct". The USDA foresees biotechnology aiding in improvement of captive management and reproduction of species, leading to more efficient species that make better use of food supplies, production of healthier **organisms**, and improvement in food and nutritional qualities of the organisms. Furthermore, aquaculture can produce organisms used as biomedical models in research, reservoirs for bioactive **molecule** pro-

duction, and organisms useful in bio remediation. Aquaculture is no longer a means of producing luxury foods, such as lobsters, but a critical solution to the world fisheries problems.

3. **Algal** aquaculture, an ancient art in Asia, not only produces seaweeds, but microalgal **cultivation** produces food supplements, such as the Omega-3 fatty acids and beta carotene. The polysaccharides of algae are a valuable commodity and a much sought after natural product. The achievements of marine biotechnology since 1983 included many important milestones. The realization that natural products associated with marine animals and plants are produced by **bacteria** is one example. It is now possible to **utilize** marine micro organisms as a source of metabolites, potential **antibiotics**, anti tumor agents, and related pharmaceuticals.

4. There is an enormous biological diversity in the world oceans, another important discovery. The Azoic theory has been disproved, a theory which held that the deepest parts of the world oceans were **devoid** of life. The discovery of the hydrothermal vents made it clear that there is an enormous diversity of life in the deep sea. The discoveries of large populations of **viruses**, of the influence of viruses in regulating algal populations, as well as of the presence of hyperthermophile viruses in the deep sea are other examples. The numbers of viruses in estuaries have been shown to be greater than that of bacteria at certain times of the year, as well as seasonality in virus **abundance** and distribution.

5. The discovery of the archaea, an ancient line of microorganisms, more closely related to higher animals and plants than to the "true bacteria", has been documented and a new determination of the **phylogeny** of life on the planet has been derived. Extremophiles, microorganisms living in extreme environments such as high pH, low pH, high hydrostatic pressure, low temperature, i. e., freezing, and extremely high temperatures(above boiling), have been isolated, characterized, and shown to be a source of commercially important products. Psychrophiles, bacteria able to grow at temperatures less than 10 °C were isolated many years ago, but the potential of psychrophiles has been underestimated. It is now clear that **enzymes** functioning at very low temperatures have valuable commercial applications. Novel species of marine bacteria have been isolated and described, including a giant bacterium visible to the naked eye. Also described, but not yet isolated, are bacteria occurring in abundance in deeper regions of the Sargasso Sea in the Atlantic Ocean and areas of the Pacific Ocean. These taxa cannot yet be grown in culture, but have been described by ribosomal RNA techniques. The use of molecular probes to discover new, as yet uncultured, taxa has been widely applied and proven exciting.

6. It has been discovered during the past decade that pure cultures are not necessarily the best means of understanding communities of bacteria in the natural environment. In fact, mixed cultures and biofilm cultures a reproving to be effective in biodegradation and in carrying out **metabolic** processes in the natural environment. Successful production of **vaccines** for aquaculture that are effective, starting in the 1970s, have provided significant advances in aquaculture of a variety of marine species, including salmonids.

7. The sea is a gigantic, largely untapped reservoir of biodiversity. Careful and cautious exploitation is essential in order not to damage and disturb this **fragile** ecosystem. The field of marine biotechnology aims to explore and utilize this biodiversity, and has great potential for beneficial outcomes for mankind. To realize this aim and potential, creative thinking and inter-disciplinary research and developments are required. The opportunity to sample marine microorganisms, including extreme thermophilic bacteria from the **geothermal vent** systems and extreme halophiles from salt ponds, can significantly expand the biomedical potential of Gulf of California organisms. The fledgling marine biotechnology industry has shown considerable interest in extreme thermophilic marine bacteria because they produce enzymes that are stable and efficient at high temperatures and pressures and are therefore attractive for use in industrial processes. The hydrothermal vent systems in the Guaymas Basin are known to be an excellent source of extreme thermophiles, but there are also many shallow-water seeps, salt ponds, **mangrove** swamps, and other unique marine microenvironments that could provide a diversity of microorganisms useful to the biotechnology industry.

    (1,043 words)

➤ *New Words*

fledgling [ˈfledʒlɪŋ]　　　　　　*a.* young and inexperienced 无经验的;刚开始的

Avatar communication has the potential to change that, and this potential defines the potential of our fledgling industry.

虚拟交流有改变的潜力,这种潜力揭示了我们这个新兴产业发展的潜力。

proximity [prɒkˈsɪmɪtɪ]　　　　*n.* the property of being close together 亲近;接近

With batch computing, the proximity of the business logic to the data significantly impacts performance.

使用批处理计算,业务逻辑与数据的接近程度将极大地影响性能。

boom [buːm]　　　　　　　　*n.* a state of economic prosperity 繁荣

Our country is basking in an economic boom.

我国正处在经济繁荣之中。

aquaculture [ˈækwəkʌltʃə]　　　*n.* rearing aquatic animals or cultivating aquatic plants for food 水产养殖;水产业

Genetics also has a key role to play in helping aquaculture meet the world's growing demand for fish.

遗传学在帮助水产养殖满足世界对鱼类日益增长的需求方面也可发挥关键作用。

organism [ˈɔːgənɪz(ə)m]　　　　*n.* a living thing that has (or can develop) the ability to act or function independently 有机体;生物体;微生物

Not all chemicals normally present in living organisms are harmless.

并非所有正常存在于活的有机体中的化学物质都是无害的。

molecule [ˈmɒlɪkjuːl]　　　　　*n.* (physics and chemistry) the simplest structural unit of an element or compound 分子;摩尔

So if we add them all up, there should be no net charge on the molecule, if the molecule is neutral.

因此如果我们把它们都加起来,这个分子上应该没有净电荷,如果这个分子是中性的话。

algal [ˈælg(ə)l]　　　　　　　*a.* relating to alga 海藻的

Similarly, changes in oceanic algal populations can, through a series of natural processes, actually lower air and water temperatures.

同样,海藻种群的改变可以通过一系列自然过程降低大气和水的温度。

cultivation [kʌltɪˈveɪʃ(ə)n]   *n.* socialization through training and education 培养;耕作;耕种

They were in the garden, but that was all; they had no share in the cultivation of its flowers.

他们虽在花园里面却不过只在花园里面罢了;他们对于花园里面花草的栽种一点也没尽过力。

bacteria [bækˈtɪərɪə]   *n.* (microbiology) single-celled or noncellular spherical organisms 细菌

When you eat off the plate, the bacteria go into your stomach.

当你吃掉了盘子里的食物时,细菌就会进入你的肚子。

utilize [ˈjuːtɪˌlaɪz]   *v.* to put into service; to make work or employ (something) for a particular purpose 利用

If applications wish to match this behavior, they must utilize and duplicate the output of those applications.

如果应用程序希望匹配这一行为,它们必须利用并复制这些应用程序的输出。

antibiotics [ˌæntɪbaɪˈɒtɪks]   *n.* a chemical substance derivable from a mold or bacterium that kills microorganisms and cures infections 抗生素;抗生学

He's hexed to discover the antibiotics.

他着了魔似的想研究出抗生素。

devoid [dɪˈvɔɪd]   *a.* completely lacking 缺乏的;全无的

How, then, could this race devoid of spirituality clothe in myths the profound horror of its life?

这个缺乏灵性的种族,那时,怎么可能为其生命深层次的恐怖披上神话的外衣呢?

virus [ˈvaɪrəs]   *n.* ultramicroscopic infectious agent that replicates itself only within cells of living hosts 病毒

The virus has infected the operating system of his computer.

病毒已使他的计算机操作系统受到感染。

abundance [əˈbʌndəns]   *n.* the property of a more than adequate quantity or supply 充裕;丰富

There is an abundance of things to talk about regarding the holidays—so go for it.

关于假期有非常丰富的事情可以讲,所以要把它提出来大家一起讨论。

phylogeny [faɪˈlɒdʒ(ə)nɪ]   *n.* (biology) the sequence of events involved in the evolutionary development of a species or taxonomic group of organisms 种系发生;系统发生

Resolution of the primate species phylogeny here provides a validated framework essential in the development, interpretation and discovery of the genetic underpinnings of human adaptation and disease.

而灵长类系统关系的解决能为发展、解释和发现人类适应和疾病的遗传基础提供合理的基本框架。

enzyme [ˈenzaɪm]　　　*n.* any of several complex proteins that are produced by cells and act as catalysts in specific biochemical reactions ［生物］酶

It is believed that it inhibits an enzyme that promotes cell proliferation in tumors.

据说它能抑制一种能在肿瘤中导致细胞增殖的酶的生成。

metabolic [metəˈbɒlɪk]　　*a.* of or relating to metabolism 变化的;新陈代谢的

So if you weigh 250 pounds then your metabolic rate will be around 2,500 calories per day.

所以,如果你体重 250 磅,那么你的新陈代谢率大约是每天 2 500 卡路里。

vaccine [ˈvækˈsiːn]　　*n.* immunogen consisting of a suspension of weakened or dead pathogenic cells injected in order to stimulate the production of antibodies 疫苗;牛痘苗

Seven million doses of vaccine are annually given to British children.

英国孩子每年要接种 700 万剂疫苗。

fragile [ˈfrædʒaɪl]　　*a.* easily broken or damaged or destroyed 脆的;易碎的

Raising prices while the world economy is in such a fragile state is "a very dangerous game", he said.

当世界经济还处在那么一个脆弱的状况,提高价格是"一个非常危险的游戏",他说。

geothermal [ˌdʒiːəʊˈθɜːm(ə)l]　*a.* of or relating to the heat in the interior of the earth 地热的;地温的

So here's what I've finally concluded: Geothermal was the right call in terms of finding an alternative to oil or gas boilers to heat the house, since we are in New England.

所以,我最后得出的结论是:既然我们住在新英格兰,那么要找到一种能够替代石油或天然气的取暖方式的话,地热是一个很不错的选择。

vent [vent]　　*n.* a hole for the escape of gas or air 出口;通风孔;(感情的)发泄

His hostility to the woman found vent in a sharp remark.

他用尖刻的话语发泄对那个女人的敌意。

mangrove [ˈmæŋɡrəʊv] *n.* a tropical tree or shrub bearing fruit that germinates while still on the tree and having numerous prop roots that eventually form an impenetrable mass and are important in land building 红树林

Hainan's mangrove and rich coral reef are well-known attractions.

海南的红树林和丰富的珊瑚礁都是著名的景观。

➤ *Phrases and Expressions*

a segment of ······的一部分
be of use 使用;利用
lead to 导致
be devoid of 缺乏
carry out 执行
a variety of 各种各样的

➤ *Proper Nouns*

European Union 欧盟
USDA (United States Department of Agriculture)美国农业部
RNA 核糖核酸
Sargasso Sea 马尾藻海(生长于大西洋,在百慕大区域)
Atlantic Ocean 大西洋
Pacific Ocean 太平洋
Guaymas Basin 瓜伊马斯盆地

➤ *Translation*

1. It may be defined as the search for commercial uses of marine biology, biochemistry... and it is a fledgling field of study having substantial potential. (Para. 1)
   它可以定义为寻找海洋生物学、生物化学和生物物理学的商业用途,是具有巨大潜力的新兴的研究领域。

2. Most of the world's tropical nations are very well suited pursue marine... to the oceans and because of their suitable climates. (Para. 1)
   世界上大多数的热带国家非常适合研究海洋生物技术,因为他们有得天独厚的条件:靠近海洋,气候适宜。
   【这个句子中有两个由 because of 组成的原因状语。翻译时如果把这种结构直接转换成汉语,译文会略显啰唆。这时,需要考虑采用增译的方法并调整语序。翻译时补充出热带地区适合研究海洋生物技术的原因,增译"得天独厚"。把原因状语调整成四字短语,并列置

于句尾,使译文表达清楚。】

3. A research collaboration between the United States and Mexico could yield considerable benefits for both countries... is experiencing a boom in biotechnology while some of the most promising locations in which to perform marine biotechnology field research are in Mexico. (Para. 1)

美国和墨西哥之间的科研合作可以为两国带来可观的利益,因为美国的生物技术正处于繁荣的发展阶段,而墨西哥地处进行海洋生物技术研究的绝佳地理位置。

4. The USDA foresees biotechnology aiding in improvement of captive management and reproduction of species... that make better use of food supplies, production of healthier organisms, and improvement in food and nutritional qualities of the organisms. (Para. 2)

美国农业部预计协助改善圈养管理和物种繁殖的生物技术,这样做能导致更多有用的物种的出现。这些物种将更有利于食品供给,带来更健康的生物,改善食品和生物的营养品质。

5. Aquaculture is no longer a means of producing luxury foods... but a critical solution to the world fisheries problems. (Para. 2)

水产养殖不再是生产诸如龙虾之类的奢侈食品的手段,而是解决世界渔业问题的关键途径。

6. It is now possible to utilize marine microorganisms as a source of metabolites... anti tumor agents, and related pharmaceuticals. (Para. 3)

现在可以利用海洋微生物作为代谢物、潜在的抗生素、抗肿瘤药物和相关药物的来源。

7. The discoveries of large populations of viruses, of the influence of viruses in regulating algal populations... hyperthermophile viruses in the deep sea are other examples. (Para. 4)

其他的发现有:大量病毒的存在,病毒在调节藻类种群方面的影响作用,以及深海热病毒的存在。

【这个句子中有三个由 of 组成的表示所属关系的名词性短语作主语。谓语处在句末。翻译时如果按原文语序结构直接翻译成汉语,译文会头重脚轻。这时,需要考虑采用调整语序的翻译方法,理顺信息量大的主语部分,调整后符合汉语的表达习惯。】

8. Novel species of marine bacteria have been isolated and described... bacterium visible to the naked eye. (Para. 5)

新物种的海洋细菌已经被分离和描述,包括肉眼可见的巨大细菌。

9. The field of marine biotechnology aims to explore and utilize this biodiversity... potential for beneficial outcomes for mankind. (Para. 7)

海洋生物技术领域的目标是探索和利用生物多样性,其具有巨大的潜力,能为人类带来有益的结果。

10. The hydrothermal vent systems in the Guaymas Basin are known to be an excellent source of extreme thermophiles, but there are also many shallow-water seeps... mangrove swamps, and other unique marine microenvironments that could provide a diversity of microorganisms

useful to the biotechnology industry. (Para. 7)

墨西哥西北瓜伊马斯盆地的地热喷口系统被认为是极端嗜热生物的极佳来源,但也有许多浅水渗流、盐池、红树林沼泽和其他独特的海洋微生物环境,这些环境可以为生物技术产业提供多种有用的微生物。

 **Exercises**

1. Fill in the blanks with the proper given words, and then translate the sentences into Chinese.

fledgling	boom	devoid	vaccine
metabolic	utilize	cultivation	abundance
unique	luxury	considerable	a variety of
ancient	extreme	represent	micro

1) We _____ 47 nations from every region of the world, and I thank each of you for being here.

2) They have warned the world to expect more frequent and intense _____ weather events, and this is what we are seeing.

3) These magnificent _____ buildings demonstrate the great intelligence of the laboring people.

4) This home-grown restaurant chain prepares _____ African game and seafood in an energetic and colourful environment.

5) And so, when you think about how much cement concrete are consumed annually on the planet, this becomes a _____ point source of greenhouse gas emissions.

6) He and his wife have a daughter and live a _____ life of some in the hills outside Oslo.

7) While each story is _____, it functions for the most part as something that can be shared and exchanged with others.

8) Things are now back to normal, but the small and _____ tourist industry has yet to recover.

9) Without this _____ in property and construction, where would the Chinese economy be in the absence of any real recovery in export markets?

10) If life was _____ of all these difficulties, we would have never understood the importance of happiness—the importance of getting what we want.

11) The _____ was developed, from start to finish, in less than a decade, in record time, and at about one-tenth of the cost usually needed to bring a product through development to the market.

12）When we go on a starvation diet to reduce our calorie intake, we not only destroy muscle tissue but also actually slow our _____ rate.

13）Listen to what they have to say and change your goals and objectives based on how your community wants to _____ social media.

14）Individuals and society will benefit from the _____ of independent self-esteem.

15）Lunar geologists and space entrepreneurs are becoming increasingly intrigued by the concept of lunar mining now that researchers have discovered a(n) _____ of water on the moon.

2. Translate the following passage into Chinese.

The sea is a gigantic, largely untapped reservoir of biodiversity. Careful and cautious exploitation is essential in order not to damage and disturb this fragile ecosystem. The field of marine biotechnology aims to explore and utilize this biodiversity, and has great potential for beneficial outcomes for mankind. To realize this aim and potential, creative thinking and inter-disciplinary research and developments are required. The opportunity to sample marine microorganisms, including extreme thermophilic bacteria from the geothermal vent systems and extreme halophiles from salt ponds, can significantly expand the biomedical potential of Gulf of California organisms. The fledgling marine biotechnology industry has shown considerable interest in extreme thermophilic marine bacteria because they produce enzymes that are stable and efficient at high temperatures and pressures and are therefore attractive for use in industrial processes. The hydrothermal vent systems in the Guaymas Basin are known to be an excellent source of extreme thermophiles, but there are also many shallow-water seeps, salt ponds, mangrove swamps, and other unique marine microenvironments that could provide a diversity of microorganisms useful to the biotechnology industry.

3. Translate the following passage into English.

海洋生物学是对海洋生物及其行为及其与环境的相互作用的研究。因为在该领域有很多话题可以研究，许多研究人员选择自己的兴趣点并专攻它。专门化研究可以基于特定的物种、生物、行为、技术或生态系统来进行。例如，海洋生物学家可能会选择研究一种蛤蜊，或者是研究当地特有蛤类。有的研究人员则参与一系列的研究活动。Alex Almario，一个实验室研究兼现场操作的技术员在网站上为科学家进行河口研究提供现场支持。他说他的多项职责包括"船的操作和维护，水质数据和样品收集，湿地、红树林、海草及珊瑚研究，实地调查和实验室生态毒理学的研究"。

Unit 2
Biotechnology and Genetic Engineering

Risk from Genetically Engineered and Modified Marine Fish

1. As the ecologically risk-free record from genetically engineered **domesticated** plants and animals lengthens, concerns of ecological risk shift to the genetic engineering of undomesticated organisms which may survive in the wild, such as fish. Released transgenic fish stocks are thought to pose a risk not only to **conspecifics**, but also to other species through niche expansion and even speciation. On the other hand, the release of hatchery fish stocks improved by classical selective breeding has attracted controversy with concern that **deleterious** hatchery **genotypes** will compromise wild genotypes.

2. Only recently, warm water marine teleosts such as the gilthead seabream have been exploited as **commercial** mariculture species, and a mixture of **molecular** and classical genetic improvement programs are underway to support this new industry. Many species of released warm water marine fish will have a great potential for **dispersal,** a low probability for **recapture** and, due to a short history of domestication, a reasonable probability of survival and reproduction in the wild. Therefore, it is timely to raise the issue of possible ecological impact from release of genetically modified warm water marine fish **prior** to extensive industrial use of genetically modified strains.

3. Here, I draw on past experiences and genetic studies of fish and terrestrial organisms to predict the likely risk from release of genetically modified marine fish. Terrestrial **invertebrates** such as Drosophila share many features in common with marine fish such as the seabream, including large population sizes, high **fecundity** and opportunity for migration. Thus from the perspective of population genetics, Drosophila offers an attractive model for the seabream.

Definition of risk

4. Among various authors, **perceptions** of ecological risk from release of genetically

modified fish vary widely. On one hand, genetically engineered changes for fish are thought to pose a risk as they might increase fitness in wild populations, give rise to niche expansion and so cause impacts not only to conspecifics but also to other members of the community. Yet classical genetic changes such as novel alleles and changes of allele frequencies usually are considered to be a risk in fish because they will be deleterious in the wild, increase genetic loads, disrupt co-adaptation, reduce genetic variation and so diminish fitness. Such **paradoxes** result, in part, from the contrasting perceptions that genetically engineered changes may be adaptive in the wild, whereas classical genetic changes usually reduce fitness in the wild, and, in part, from the use of different value judgments about what is beneficial or harmful for the population or environment.

5.   For this paper, a **consensus** definition may be reached by defining ecological risk as having the potential to occur whenever genetic changes resulting from human activities have a non-**negligible** probability of increasing in the wild. Genetic changes (hereafter referred to as laboratory genetic changes) include changes of allele and genotype frequencies, novel alleles, chromosome mutations, hybrids, genetically-engineered changes and newly formed species. Negligible probability events are those considered too unlikely to warrant **precaution**, such as the event of a meteorite striking a given structure, one in 500,000 year events, etc. The "wild" is taken to be an environment without new and substantial selection pressures generated by man, but with an established population of conspecifics to **facilitate** intra-specific competition.

6.   Thus my definition of risk is restrictive in that it does not apply to environments that have been substantially disturbed by man through widespread introduction of antibiotics, insecticides, fertilizers, herbicides, pollution, and **exotic** species including new **pathogens.** These environments are not uncommon for either terrestrial or marine ecosystems. Should conspecifics survive in disturbed environments, new selection pressures generated indirectly by man may operate on new or existing genetic variance from either the wild or the laboratory, in accordance with relative abundance, extent of genetic variance and probability for new mutation. Also, environments with depleted populations of conspecifics may relax selection pressures against laboratory genetic changes. For example, absence of conspecific or related species such as wild wolves or pigs may allow domesticated dogs and pigs to become feral. Issues of whether the primary risk is the **disturbance** itself, or the laboratory genetic changes which survive in disturbed environments, and options to moderate such impact are not considered here.

7.   Also, my definition does not consider the problems of exotic disease transmission to wild strains through transport and release of cultured fish strains, as these risks arise

primarily from husbandry considerations, such as the **propagation** and transport of pathogens with cultured fish strains. Genetic differences between wild and cultured fish strains would be relevant for the release of an exotic pathogen along with a genetically resistant cultured strain (either classically selected or genetically-engineered), although the **ubiquitous** presence of an exotic pathogen results by definition in a disturbed environment.

Factors contributing to potential risk

8. Laboratory genetic changes with deleterious effects on fitness can increase in the wild principally from large scale and repeated releases. Stochastic factors such as drift should be of less immediate importance in view of the large natural populations for most maricultured fish. Also, theoretically, genetic process such as different types of drive could increase deleterious changes. Drive elements in fish have not been detected, and inheritance patterns of qualitative phenotypes in fish suggest they are rare. Moreover, Drosophila studies indicate the likely rapid evolution of suppressor modifiers of meiotic drive especially for deleterious genetic changes, although meiotic drive factors in mice have persisted through long periods of time.

9. On the other hand, should laboratory genetic changes confer fitness advantages, then natural selection of the change, either as new adaptive intra-specific **polymorphism** or as a component of a new species, may increase the frequency of the change in the wild despite low levels of release.

10. Thus here different genetic improvement programs whether classical or molecular are assessed for potential risk using three principal criteria, namely a) the magnitude of planned or accidental releases, b) whether the genetic change will be selected in the wild, and c) whether the genetic change will yield a new species distinct from the parental species through reproductive or ecological isolation.

(1,106 words)

➤ *New Words*

domesticated [dəˈmestɪˌkeɪtɪd]   *a.* converted or adapted to domestic use 家养的；驯养的
RVF（Rift Valley Fever）is able to infect many species of animals causing severe disease in domesticated animals including cattle, sheep, camels and goats.
裂谷热能够感染许多动物,使包括牛、羊、骆驼和山羊在内的家养动物患严重疾病。

conspecific [kɒnspəˈsɪfɪk]   *n.* an organism belonging to the same species as another organism 同种个体
Spawning is very inconspicuous—no shaking or sparring with conspecifics.
繁殖过程很隐蔽,雄鱼不抖动,也不追咬同种。

deleterious [ˌdeləˈtɪərɪəs]   *a.* harmful to living things 有毒的；有害的
However, most mutations are deleterious.
然而,大多数突变是有害的。

genotype [ˈdʒenətaɪp]   *n.* a group of organisms sharing a specific genetic constitution 基因型
The large majority of units produce either meat or eggs, with specialized broiler or layer genotypes.
绝大部分养殖场生产具有特定肉鸡或者下蛋鸡基因型的禽肉或禽蛋。

commercial [kəˈmɜːʃ(ə)l]   *a.* of or relating to commercialism 商业的；盈利的
Our correspondence is limited to a few commercial letters.
我们之间的通信只限于几封商业函件。

molecular [məˈlekjʊlə]   *a.* relating to or produced by or consisting of molecules 分子的；由分子组成的
You do not get this line from molecular hydrogen.
从氢分子中我们得不到这种光线。

dispersal [dɪˈspɜːsl]   *n.* the process of spreading things over a wide area or in different directions 散布；分散；传播
Plants have different mechanisms of dispersal for their spores.
植物传播孢子的方法各不相同。

recapture [riːˈkæptʃə]   *v.* to bring back the same feelings or qualities that you experienced in the past 再体验；夺回；拿回
But I cannot recapture its beauty.

但是我却不能再次体验到它的美。

prior ['praɪə]

a. existing or arranged before something else or before the present situation 优先的；在前的

It happened prior to his arrival.

这事发生在他到达之前。

invertebrate [ɪn'vɜːtɪbrət]

a. lacking a backbone or spinal column 无脊椎的；无骨气的

Many species of invertebrate have shells or skeletons made of calcium carbonate.

好多无脊椎类的物种有碳酸钙的外壳或骨骼。

fecundity [fɪ'kʌndətɪ]

n. the intellectual productivity of a creative 繁殖力；多产；肥沃

In a world where everything is given and nothing is explained, the fecundity of a value or of a metaphysic is a notion devoid of meaning.

在一个万物皆已给出而却不加以解释的世界中，价值或形而上学的多产性只是个意义尽失的概念。

perception [pə'sepʃ(ə)n]

n. the representation of what is perceived 知觉；感觉；看法

When they do, what do you think their perception will be?

当他们这样做时，你觉得他们的知觉是怎么样的？

paradox ['pærədɒks]

n. a statement that contradicts itself 悖论

Is this a paradox of faith?

这是个信仰的悖论吗？

consensus [kən'sensəs]

n. agreement in the judgement or opinion reached by a group as a whole 共识；一致

There was no consensus as to what shape reforms should take.

对于应当采取何种形式的改革，并没有一致意见。

negligible ['neglɪdʒəb(ə)l]

a. so small as to be meaningless; insignificant 微不足道的；可以忽略的

The cost and complexity for each deployment become negligible.

每个部署的成本和复杂性变得微不足道。

precaution [prɪ'kɔːʃ(ə)n]

n. a measure warding off impending danger 预防；警惕

Could he not move to a place of safety, just as a precaution.

难道他就不能为了防范搬到一个安全的地方去吗？

facilitate [fə'sɪlɪteɪt]

v. to make easier 促进；帮助；使容易

The new airport will facilitate the development of

tourism.

新机场将促进旅游业的发展。

exotic [ɪˈgzɒtɪk]   *a.* being or from or characteristic of another place 异国的;外来的

Find out which of these are native plants and which are exotic.

找出其中哪些是原生植物,哪些是外来植物。

pathogen [ˈpæθədʒ(ə)n]   *n.* any disease producing agent 病原体;病菌

The reservoir of this pathogen appears to be mainly cattle and other ruminants such as camels.

这一病菌的宿主主要是家禽和其他反刍动物,例如骆驼。

disturbance [dɪˈstɜːb(ə)ns]   *n.* activity that is an intrusion or interruption 干扰;骚乱

It's impossible to localize the effect of this political disturbance.

无法将这场政治骚乱的影响限制在局部范围内。

propagation [ˌprɒpəˈgeɪʃ(ə)n]   *n.* the spreading of something (a belief or practice) into new regions 传播;繁殖

That requires another step: propagation of changes to the project models from the enterprise models.

这就需要另一步:从企业模型中传播变更到项目模型中。

ubiquitous [juːˈbɪkwɪtəs]   *a.* being present everywhere at once 普遍存在的;无所不在的

We are gradually beginning to realize that the potential for life is ubiquitous throughout our galaxy.

我们慢慢开始意识到在我们的银河系中可能普遍存在生命。

polymorphism [ˌpɒlɪˈmɔːfɪz(ə)m] *n.* the occurrence of more than one form of individual in a single species within an interbreeding population 多态性

So long as method names match, you have polymorphism.

只要方法名匹配,就可以利用多态性。

➤ *Phrases and Expressions*

concern of 关注
the release of 释放
be exploited as 被利用为
due to 由于;归因为
prior to 在……之前
in accordance with 依照;与……一致
apply to 适用于;应用于
contribute to 有助于;捐献

➤ *Translation*

1. As the ecologically risk-free record from genetically engineered domesticated plants and animals lengthens, concerns of ecological risk shift to the genetic engineering of... which may survive in the wild, such as fish. (Para. 1)

   由于基因工程培养植物和动物生态无风险的记录延长,人们对生态风险的相关关注转移到可能在野外生存的野生生物,如鱼类的基因工程上。

2. Many species of released warm water marine fish will have a great potential for dispersal, a low probability for recapture and, due to a short history of domestication... of survival and reproduction in the wild. (Para. 2)

   许多温水海洋鱼类的传播潜力巨大,被再次捕获的可能性小,因为它们养殖史短,野外生存和繁殖的可能性高。

3. Terrestrial invertebrates such as Drosophila share many features in common with marine fish such as the seabream..., high fecundity and opportunity for migration. (Para. 3)

   果蝇这样的陆地无脊椎动物与鱼类有很多相似之处。鲷鱼也是,其数量多、繁殖力强、迁徙的机会多。

4. On one hand, genetically engineered changes for fish are thought to pose a risk as they might increase fitness in wild populations... and so cause impacts not only to conspecifics but also to other members of the community. (Para. 4)

   一方面,鱼的基因改变被认为是一种风险,因为它们可能会提升野生种群的适应性,引起生态位的扩张,因此造成的影响不仅是针对同种个体,也是针对整个种群的。

5. Yet classical genetic changes such as novel alleles and changes of allele frequencies usually are considered to be a risk in fish because they will be deleterious in the wild... disrupt co-adaptation, reduce genetic variation and so diminish fitness. (Para. 4)

   然而,诸如新的等位基因和等位基因频率变化的传统基因改变也是有风险的。通常认为基因改变的鱼类在野生环境中生存有缺陷,因为这样会增加遗传荷载,破坏共生性,减少基因变异,因此降低适应性。

【这个句子是含有原因状语从句的复合句。翻译时按照先果后因的语序。在被动句的处理上转换成汉语中的主动句。找出句子的核心内容是"基因改变是有风险的"。最后按照顺序的译法把原因状语从句中要表述的原因一一解释清楚。】

6. The "wild" is taken to be an environment without new and substantial selection pressures generated by man... of conspecifics to facilitate intra-specific competition. (Para. 5)

"野生条件"不是人类人为设置的一个环境,没有新的和巨大的选择压力,而是特定的同种群体环境,以促进种内竞争。

7. Should conspecifics survive in disturbed environments, new selection pressures generated indirectly by man may operate on new or existing genetic variance... in accordance with relative abundance, extent of genetic variance and probability for new mutation. (Para. 6)

根据相对的丰富程度、基因变异的程度和新突变的可能性而言,不管同种个体是否应该在干扰下的环境中生存,人类间接带来的新的选择压力或许会对新的或现有的遗传变异产生影响。这些基因变异在野生条件下或者在实验室里进行。

【这个句子结构复杂。在翻译时通读原文,做到信、达、雅。使读者理解即为信;用通顺的文本转述原文即为达;文字优美即为雅。翻译时为了让整个句子通顺,调整了句子的语序,把句尾的介词短语内容调至句首,同时对"either...or..."的隐含内容进行补译。】

8. Issues of whether the primary risk is the disturbance itself, or the laboratory genetic changes which survive in disturbed environments... such impact are not considered here. (Para. 6)

这里没有考虑诸如主要风险是干扰本身的问题,还是在受干扰的环境中生存的实验室基因变化问题,以及对这种影响的调整选择问题。

9. Genetic differences between wild and cultured fish strains would be relevant for the release of an exotic pathogen along with a genetically resistant cultured strain (either classically selected or genetically-engineered)... of an exotic pathogen results by definition in a disturbed environment. (Para. 7)

野生和人工养殖的鱼种之间的遗传差异与一种外来病原体的释放以及具有遗传抗性的培养菌株(传统选择或基因工程)有关,尽管从概念上来说受干扰的环境中普遍存在一种外来病原体。

10. Laboratory genetic changes with deleterious effects on fitness can increase... from large scale and repeated releases. (Para. 8)

实验室的基因改变对鱼类会产生危害,主要是由于大规模和反复放养造成的。

 **Exercises**

1. Fill in the blanks with the proper given words, and then translate the sentences into Chinese.

controversy	deleterious	recapture	prior
perception	precaution	facilitate	exotic
disturbance	ubiquitous	exploit	substantial
relevant	indicate	yield	ecological

1) _____ balance would be broken by the fact that rabbits over abound on the grasslands.

2) The people who were held down ruthlessly would not _____ to aggressors.

3) The method does not _____ which species, or how many mosquitoes, deposited viruses on the cards.

4) I will read anything they write, because I know it will be both _____ to me and an enjoyable reading experience.

5) This sounds promising, but as far as we know, _____ investigation of the applicable rules has yet to be done.

6) Publishers from across the globe have gathered in London this week to discuss how to _____ this growing opportunity.

7) But previous studies have linked noise to heart attacks and higher blood pressure, too—possibly due to increased stress and sleep _____.

8) If you just want to add some shine to your resume, or taste the _____ life, it's not worth the cost.

9) Because the need for global cooperation is becoming more evident, the need for creating global classrooms to _____ such collaboration is apparent as well.

10) It may surprise many to know that the dairy farm environment, even when every _____ is taken, is a reservoir for illness-causing germs.

11) Since the past is the past, and we can't go back and change it, is it possible to change our reality and our _____ of the world?

12) If it doesn't work properly, you should then work with the appropriate network teams to resolve the problems _____ to deployment.

13) Obama seeks to _____ some momentum of his own after weeks of being on the defensive over his relationship with his former pastor, Wright.

14) Many slightly beneficial mutations can be lost by chance, while mildly _____ ones can spread and become fixed in a population.

15) Global warming and climate change have been the focus of _____ and debate for nearly

half a century—Is global warming real?

2. Translate the following passage into Chinese.

Among various authors, perceptions of ecological risk from release of genetically modified fish vary widely. On one hand, genetically engineered changes for fish are thought to pose a risk as they might increase fitness in wild populations, give rise to niche expansion and so cause impacts not only to conspecifics but also to other members of the community. Yet classical genetic changes such as novel alleles and changes of allele frequencies usually are considered to be a risk in fish because they will be deleterious in the wild, increase genetic loads, disrupt co-adaptation, reduce genetic variation and so diminish fitness. Such paradoxes result, in part, from the contrasting perceptions that genetically engineered changes may be adaptive in the wild, whereas classical genetic changes usually reduce fitness in the wild, and, in part, from the use of different value judgments about what is beneficial or harmful for the population or environment.

3. Translate the following passage into English.

基因工程生物是在实验室中测试和生成人类所需的生物品质。最常见的做法是将一个或多个基因添加到生物体的基因组中。少数做法是基因被移除,或是基因表达被增强,或是基因失效,又或是增、减基因复制的数量。研究过程中一旦出现令人满意的基因,研究者就会申请监管批准,进行实地测试,这被称为"实地释放"。实地测试就是在一个受控的环境中培育农场里的植物,或者孵化动物。如果这些现场测试成功,生产者就会申请监管批准,来种植和销售农作物。

Unit 3

Applications of Marine Creatures

Seahorses—A Source of Traditional Medicine

1. Intensive research since 1970s has proved that marine organisms are magnificent sources of bioactive compounds. Marine organisms are collected for the discovery and development of **pharmaceutical** drugs used in allopathic medicine (western medicine, evidence based medicine, and biomedicine). In recent years, many bioactive compounds have been extracted from various marine animals such as tunicates, sponges, soft corals, sea hares, nudibranchs, bryozoans, sea slugs and marine organisms. Extracts from the collected organisms are tested for their effectiveness against particular disease targets in a series of **automated** screens. If active, the compound responsible is isolated and its molecular structure is determined by secondary testing done on **efficacy** before the decision is made to subject the compound to **preclinical** and possibly **clinical** trials. Very few compounds succeed in becoming commercial products; the process can take 10–15 years (or longer) and can cost hundreds of millions of dollars. Whilst tens (if not hundreds) of thousands of marine species have likely been sampled, only about 20 marine compounds are currently in clinical **trials**. Removal of marine organisms to supply this process can be broken into two types: primary collections and secondary or re-collections. Primary collections are typically broad and **speculative** in order to maximise the possibility of discovering bioactive compounds during screening, whereas secondary re-collections are focused on supplying a particular "bioactive" species of interest to later tests during the drug development. Hence, this article emphasizes the biomedical properties of seahorses in a review aspect.

2. Seahorses are bony fish belonging to the family Syngnathidae. The family Syngnathidae also includes pipefishes, pipehorses and seadragons. The primary taxonomic groupings within this family reflect the location and development of the male brood pouch, head/body axis, development of the caudal fin, and **prehensile** ability of the tail. All seahorses belong to one genus, Hippocampus. According to the report, there are 32

known species of seahorses present world wide. Seahorses are found in both temperate and tropical shallow coastal waters (5,150 m depth), with a latitudinal distribution from about 50 north to 50 south and with the greatest species diversity in the Indo-Pacific.

3. Traditional medicine or **complementary** and alternative medicine is important to the health care of millions of people worldwide. TM comes in at least 125 recognized forms, including the **codified** systems of traditional Chinese medicine (TCM), Ayurveda, Unani and the unwritten "folk" medicines of the Americas, sub-Saharan Africa and the Asia-Pacific region. The majority of these were consumed in Asia, representing the practices of TCM, Hanyak, Kanpo, Ayurverda, Unani, Jamu and other folk medicines, which is an absolute global minimum, and identified several areas absent from the analysis such as the Middle East and North Africa. Relatively little is known about the status of many marine medicinal populations or indeed the size and biological significance of the TM component of **mortality.** Due to its **vulnerability**, 23 species of seahorses were listed in either Appendix I or II of the Convention of International Trade in Endangered Species. Seahorses have been used in TCM for a long time in various forms, such as dried seahorses.

4. As a medicinal source, seahorses are used as an **ingredient** in traditional medicine, particularly in Southeast Asia where traditional Chinese medicine and its derivatives are practised and have been used perhaps for about 600 years. Seahorses are credited with having a role in increasing and balancing vital energy flows within the body, as well as a **curative** role for ailments such as impotence and **infertility**, asthma, high cholesterol, goiter, kidney disorders, and skin afflictions such as severe acne and persistent nodules. They are also reported to **facilitate** parturition and act as a powerful genital tonic and as a potent aphrodisiac. In Brazil, whole seahorses are an important medicinal resource used to treat asthma and gastritis. According to Vincent , the Chinese generally regard the historical use of TCM as a **testimony** to a product's efficacy, and clinical trials are rare; but there have been 10 publications relating to medicinal properties such as anti-ageing effects and improved **immune** responses. In the Philippines aquarium trade, "yellow" seahorses are usually more valuable than "black" seahorses. Until at least the eighteenth century, seahorses were also **utilized** for their medicinal properties in many western countries, with applications recorded back to 342 B. C. where they were reputed to have medicinal properties with regard to baldness, leprosy, urine retention and rabies.

5. Traditional Chinese medicine is recognized by the WHO as a valid form of medicine and is accepted by more than one-quarter of the world's population. Dried seahorse species that are **predominantly** used in TCM are Hippocampus histrix. Seahorses

proved to be a source of controlling aging process by possessing **immense** antioxidants that are much **evaluated** from the seahorse species of H. kuda Bleeler. The report shows that seahorses play a vital role in **scavenging** activities in ageing phenomenon. In addition to these, seahorses have an immense anticancer, antifatigue compound in them. Recently, an indepth molecular analysis of an anticancer compound from seahorses was identified by **synthesizing** a novel phthalate derivative from the seahorse, which immensely inhibits the Cathepsin, which is responsible for the cause of many cancer cell growth and neurogenerative disorders. In addition to this, an innovative class of anti-inflammatory peptide, isolated from the seahorse H. kuda, exhibits characteristic effects against arthritis, which is a major chondrocytic degenerative disease characterized by the **degradation** of articular cartilage involving excessive degradation of extracellular matrix (ECM) and synovial inflammation. A nutritive analysis and **proximate** composition of six seahorse species was done from the Chinese coast, and it reveals that the seahorse contains medically important nutrients in the form of proteins and poly unsaturated fatty acids that prevent cardiac problems. Apart from this, seahorse shavean efficacy of antimicrobial potence against some dreadful microorganisms, especially against Klebsilla pneumonia. Hence, the traditional system of seahorse consumption is **authentically** proved to have a biomedical value highly supported by this research outcome.

6. Seahorses are a leading drug candidate species for numerous diseases. They are a pharmacological mine for various diseases such as cancer and impotence. Hence, a broad-based analysis of seahorses should be done for evaluating the biomedical potential to come up with an active drug compound from them.

(1,040 words)

➤ *New Words*

pharmaceutical [ˌfɑːməˈsuːtɪk(ə)l]　*a.* of or related to pharmacy or pharmacists 制药的;配药的
She's a biologist for a pharmaceutical company.
她是一家制药公司的生物学家。

automated [ˈɔːtəmeɪtɪd]　*a.* operated by automation 自动化的;机械化的
But how do they verify this in an automated environment?
但是,在自动化环境中如何检查这一点呢?

efficacy [ˈefɪkəsɪ]　*n.* capacity or power to produce a desired effect 功效;效力
How do you assess the utility and efficacy of this channel in today's environment?
你如何评价这一渠道在当今环境中的实用性和功效?

preclinical [priːˈklɪnɪk(ə)l]　*a.* of or related to the early phases of a disease when accurate diagnosis is not possible because symptoms of the disease have not yet appeared 临床前的;现出症状之前的潜伏期的
He added that the vaccine had also demonstrated safety in preclinical toxicology studies.
他补充说,这种疫苗已在临床前毒理学研究中被证明是安全的。

clinical [ˈklɪnɪk(ə)l]　*a.* relating to a clinic or conducted in or as if in a clinic and depending on direct observation of patients 临床的,诊所的
At present we do not have programs for clinical psychology.
目前,我们没有任何关于临床心理学的规范。

trial [ˈtraɪəl]　*n.* legal proceeding consisting of the judicial examination of issues by a competent tribunal 试验;审讯
So far this is the only trial of the treatment.
迄今为止,这是该疗法唯一的试验。

speculative [ˈspekjələtɪv]　*a.* not based on factor investigation 推测的
He has written a speculative biography of Christopher Marlowe.
他写了一篇关于克里斯托弗·马洛的推测性传记。

prehensile [prɪˈhensaɪl]　*a.* (of a part of an animal's body 动物肢体的一部分) able to hold things 能抓住东西的;缠绕性的
Long and slightly prehensile, but more like the tail of a

panther than a monkey.

尾巴是细长而且可缠绕的,但比起猴子那更像猎豹的尾巴。

complementary [ˌkɒmplɪˈmentərɪ]   *a.* of words or propositions so related that each is the negation of the other 补足的,补充的

That would be good for my apprentice to have, complementary.

补充说一下,那会是成为我徒弟的优势。

codify [ˈkəʊdɪfaɪ]   *v.* to organize into a code or system, such a body of law 编纂

Ultimately all we have is those systematic laws that codify the way things behave.

最终我们对所有事物行为的编纂归类都基于其系统法则。

mortality [mɔːˈtælətɪ]   *n.* the quality or state of being mortal 死亡率

Infant mortality in Palestine refugees is among the lowest in the Near East.

巴基斯坦难民中的婴儿死亡率在近东地区是最低的。

vulnerability [ˌvʌlnərəˈbɪlətɪ]   *n.* the state of being vulnerable or exposed 易损性;弱点

This points to another vulnerability.

这指的是另一个弱点。

ingredient [ɪnˈgriːdɪənt]   *n.* a component of a mixture or compound 原料;要素;组成部分

What is the secret ingredient that keeps some restaurants in steady business for decades?

维持一些餐厅数十年来稳定经营的秘密要素是什么呢?

curative [ˈkjʊərətɪv]   *a.* tending to cure or restore to health 有疗效的;治病的

This means, in ethical terms, that no one in need of health care, whether curative or preventive, should risk financial ruin as a result of having to pay for care.

这意味着,在伦理上,需要卫生保健服务的人,不管其需要的是治疗,还是预防服务,都不得因支付卫生保健费用而陷入经济困境。

infertility [ˌɪnfɜːˈtɪlətɪ]   *n.* the state of being unable to produce offspring 不育

Currently, these programs do not educate people on the link with prevention of infertility.

目前,这些规划未向人们阐明其与预防不孕症之间的关系。

facilitate [fəˈsɪlɪteɪt]　　　　　*v.* to make easier 促进；帮助；使容易
　　　　　　　　　　　　　　　You could facilitate the process by sharing your knowledge.
　　　　　　　　　　　　　　　你可以通过分享知识促进进程。

testimony [ˈtestɪmənɪ]　　　　*n.* a solemn statement made under oath 证词；证据
　　　　　　　　　　　　　　　The testimony of witnesses vindicated the defendant.
　　　　　　　　　　　　　　　证人们的证词证明了被告的无辜。

immune [ɪˈmjuːn]　　　　　　*a.* a person who is immune to a particular infection 免疫的
　　　　　　　　　　　　　　　The immune system is our main defence against disease.
　　　　　　　　　　　　　　　人体的免疫系统是抵御疾病的主要屏障。

utilize [ˈjuːtəlaɪz]　　　　　　*v.* to put into service；make work or employ for a particular
　　　　　　　　　　　　　　　purpose 利用
　　　　　　　　　　　　　　　We must utilize all available resources.
　　　　　　　　　　　　　　　我们必须利用可以得到的一切资源。

predominantly [prɪˈdɒmɪnəntlɪ]　*ad.* much greater in number or influence 主要地；显著地
　　　　　　　　　　　　　　　Those are the reasons predominantly why people cannot
　　　　　　　　　　　　　　　get out themselves.
　　　　　　　　　　　　　　　这就是人们为什么无法撤出的主要原因。

immense [ɪˈmens]　　　　　　*a.* unusually great in size or amount or degree or especially
　　　　　　　　　　　　　　　extent or scope 巨大的；无边无际的
　　　　　　　　　　　　　　　One disease on our agenda causes immense suffering in
　　　　　　　　　　　　　　　large parts of the world，but does its greatest harm in
　　　　　　　　　　　　　　　Africa.
　　　　　　　　　　　　　　　我们的议程上有一种疾病在世界上许多地方造成巨大
　　　　　　　　　　　　　　　的痛苦，在非洲的危害最大。

evaluate [ɪˈvæljʊeɪt]　　　　　*v.* to from an opinion of the amount，value or quality of
　　　　　　　　　　　　　　　something after thinking about it carefully 估价；评价；
　　　　　　　　　　　　　　　评估
　　　　　　　　　　　　　　　Don't evaluate people by their clothes.
　　　　　　　　　　　　　　　不要根据穿着来评价人。

scavenge [ˈskævɪndʒ]　　　　*v.* to clean refuse from 打扫；排除废气
　　　　　　　　　　　　　　　You can often scavenge nice bit of old furniture from
　　　　　　　　　　　　　　　skip.
　　　　　　　　　　　　　　　从废物堆里往往能捡到一些挺好的旧家具。

synthesize [ˈsɪnθəsaɪz]　　　　*v.* to combine so as to form a more complex product 合成；
　　　　　　　　　　　　　　　综合
　　　　　　　　　　　　　　　There are many amino acids that the body cannot synthesize itself.
　　　　　　　　　　　　　　　有许多种氨基酸人体自身不能合成。

degradation [ˌdegrəˈdeɪʃ(ə)n]	*n.*	changing to lower state 退化;降格
		What is land degradation?
		什么是土地退化?
proximate [ˈprɒksɪmət]	*a.*	closest in degree or order especially in a chain of causes and effects 近似的;最近的
		In nearly every start-up that fails, the proximate cause is running out of money.
		几乎每场失败的创业都是直接因为资金断档造成的。
authentically [ɒˈθentɪklɪ]	*ad.*	genuinely; with authority 真正地;确实地;可靠地
		In fact, we can only love others authentically when we love ourselves.
		事实上,只有当我们爱自己时,我们才能真正地爱他人。

➤ *Phrases and Expressions*

extract from 从……提取
be determined by 由……所决定
succeed in 在……方面成功
focus on 集中于……
the majority of ……的大多数
play a role in 在……起作用
come up with 提出;想出

➤ *Proper Names*

TM 传统药物
TCM 传统中药
Middle East 中东地区
North Africa 北非
Convention of International Trade in Endangered Species 濒危物种国际贸易公约
WHO 世界卫生组织

➤ *Translation*

1. In recent years, many bioactive compounds have been extracted from various marine animals... nudibranchs, bryozoans, sea slugs and marine organisms. (Para. 1)
   近年来,许多生物活性化合物从各种海洋动物中提取出来,如被囊类动物、海绵、软珊瑚、海兔、裸鳃鱼、苔藓虫、海蛞蝓和海洋有机生物。

2. Extracts from the collected organisms are tested for their effectiveness against particular disease

targets in a series of automated screens. (Para. 1)

对收集到的生物体中的提取物进行了有效性测试,尤其针对抵抗特定的疾病进行的一系列自动筛查。

3. Primary collections are typically broad and speculative in order to maximise the possibility of discovering bioactive compounds during screening... are focused on supplying a particular "bioactive" species of interest to later tests during the drug development. (Para. 1)

最初的收集通常是广泛性和推测性的,目的是在筛选过程中最大限度地发现生物活性化合物的可能性,而二次收集的重点是在药物开发过程中提供一个特定的"生物活性"物种,以进行后期的测试。

4. Seahorses are found in both temperate and tropical shallow coastal waters (5, 150 m depth), with a latitudinal distribution from about 50 north to 50 south and with the greatest species diversity in the Indo-Pacific. (Para. 2)

海马是印度洋-太平洋中,跨越南北纬50度,在温带和热带浅水区域(5 150米深)都有分布的最多的物种。

【这个句子中翻译的难点是怎样处理独立主格结构。原文中分别交代了海马生活的地理位置、气候条件。如果按照原文的语序直译,译文会略显啰唆。译文按照海马生活的地理位置的大小、范围,将原句的语序调整,直接翻译成符合汉语习惯的主语+谓语结构,清晰明了。】

5. Relatively little is known about the status of many marine medicinal populations or indeed the size and biological significance of the TM component of mortality. (Para. 3)

对于许多海洋药物种群的地位,以及传统药物死亡率的高低和生物学意义,我们知之甚少。

6. As a medicinal source, seahorses are used as an ingredient in traditional medicine, particularly in Southeast Asia where traditional Chinese medicine... have been used perhaps for about 600 years. (Para. 4)

海马作为传统医学成分的药用来源,在沿用中国传统药物及衍生物的东南亚国家被研究使用了六百多年。

7. Until at least the eighteenth century, seahorses were also utilized for their medicinal properties in many western countries, with applications recorded back to 342 B. C. ... have medicinal properties with regard to baldness, leprosy, urine retention and rabies. (Para. 4)

至少在18世纪以前,许多西方国家利用海马的药用特性。这些应用可以追溯到公元前342年,当时人们认为海马具有治疗秃顶、麻风病、尿潴留和狂犬病的药用价值。

8. Recently, an indepth molecular analysis of an anticancer compound from seahorses was identified by synthesising a novel phthalate derivative from the seahorse... which is responsible for the cause of many cancer cell growth and neurogenerative disorders. (Para. 5)

最近,人们对来自海马的抗癌化合物进行了深入的分子分析,通过合成法得到了来自海马的衍生物邻苯二甲酸盐。它极大地抑制了组织蛋白酶的生长,这种组织蛋白酶是引发许多癌细胞生长和神经退行性疾病。

【这个句子是长难句。翻译时首先把专业术语理顺。其次通读原文,分清主谓关系。再次采用拆译的方法层层剥离。主句是... was identified by... 的结构,随后是由两个"which"引导的非限定性定语从句,最后增译了主语"人们"使翻译表达清楚。】

9. A nutritive analysis and proximate composition of six seahorse species was done from the Chinese coast... contains medically important nutrients in the form of proteins and poly unsaturated fatty acids that prevent cardiac problems. (Para. 5)

从中国沿海地区取样,对 6 种海马物种进行了营养分析和近似成分分析,发现海马体内含有蛋白质和多不饱和脂肪酸的重要营养成分,它们可以预防心脏疾病。

 **Exercises**

1. Fill in the blanks with the proper given words, and then translate the sentences into Chinese.

automated	clinical	trial	prehensile
complementary	mortality	vulnerability	ingredient
infertility	facilitate	utilize	immense
evaluate	degradation	proximate	authentically

1) However, this hypothesis is a _____ explanation and is perfectly compatible with the ultimate explanation we have focused on here.

2) The aim of the field experiment was to investigate to what extent such a system could prevent soil _____ resulting from very intensive crop rotations.

3) By now to _____ them is how to tell whether they are good arguments or bad arguments.

4) Clearly, I put him this high on the list due to his _____ fame rather than his stunning knowledge of football talent.

5) Knowing the people you are working with and their personality types can help you _____ their strengths and overcome the challenges ahead to a more successful journey.

6) Because the need for global cooperation is becoming more evident with every succeeding year, the need for creating global classrooms to _____ such collaboration is apparent as well.

7) In the effort to be distanced from being what you don't want to become, you try to become you _____.

8) What is the secret _____ that keeps some restaurants in steady business for decades?

9) There is also, of course, the issue of how to do the testing, especially the extent to which tests should be _____.

10) However, researchers and pharmaceutical manufacturers have significantly cut back on funds

for the evaluation and _____ testing of new agents.

11) As a scientist I am not happy with just one _____ and there will be a large debate.

12) They can do this on the ground, but also in trees due to their long claws and strong _____ tails.

13) If you have done your research and wish to try a _____ therapy, the next step is to talk to your personal doctor or HIV specialist.

14) An investment in safe water and sanitation for homes and schools can be a key factor in reducing child _____.

15) You not only empower other people, you also encourage them to take on additional responsibilities and establish a bond of trust based on your _____ and their capability.

2. Translate the following passage into Chinese.

Seahorses are bony fish belonging to the family Syngnathidae. The family Syngnathidae also includes pipefishes, pipehorses and seadragons. The primary taxonomic groupings within this family reflect the location and development of the male brood pouch, head/body axis, development of the caudal fin, and prehensile ability of the tail. All seahorses belong to one genus, Hippocampus. According to the report, there are 32 known species of seahorses present world wide. Seahorses are found in both temperate and tropical shallow coastal waters (5,150 m depth), with a latitudinal distribution from about 50 north to 50 south and with the greatest species diversity in the Indo-Pacific.

3. Translate the following passage into English.

到哪里寻找有趣的潜在的药物? 可以从具有高度生物多样性的地方开始,比如从珊瑚礁、海岸红树林、海底山或其他生物丰富的区域入手。在那里有可能会出现许多物种和有趣的化学物质。也可以从像海绵、珊瑚、被囊类等固着动物入手,因为它们数量多,种类丰富,易捕获。这些动物主要通过化学手段进行防御或保护自己。生活在它们体内的许多共生细菌也能产生有用的化学物质。另外,极端条件下的一些特别的地方,例如深海热泉和北极地区,可能也会给人类带来生物学上的惊喜。尽管在上述地区取样难度要大得多,但是最终产品的应用除了效力于生物制药外也可能会更广泛。

Chapter 2

Fishery Resources and Environment

Unit 1
Sustainable Development of Fishery Resources

Transitioning to Sustainable Fisheries Management

1. What do people feel needs to be done to ensure that we have at least some chance of keeping our marine fisheries productive, as well as identifying examples of good practice around the world that could be **scalable** for all our benefit, asks a report by the International Sustainability Unit (ISU).

2. The increased pressure on marine capture fisheries—from growing populations, rising demand for seafood, and a rapid increase in fisheries exploitation—has caused a decline in the productivity of many fisheries. According to the UN Food and Agriculture Organization (FAO), 32 percent of fish stocks are now overexploited, **depleted** or recovering from depletion, and this figure is rising every year. As well as delivering less food and income and supporting fewer livelihoods than they could, overexploited fisheries are also more vulnerable to external pressures such as climate change and pollution. However, research and consultation undertaken by the ISU shows that practical solutions are available.

3. This report presents a synthesis of research commissioned by the ISU, together with findings from a broad stakeholder consultation. It seeks to outline the critical importance of wild fish stocks and the benefits that come from their sustainable management. It shows how these benefits are already being realized in fisheries around the world through the implementation of a wide range of tried and tested tools for sustainability. If managed responsibly, wild fish stocks can play a crucial role in food security, sustainable livelihoods and **resilient** economies.

4. Wild fish stocks are of enormous importance to economic output, livelihoods and food security. If degraded fisheries are rebuilt and sustainably managed, they can make an even larger contribution. The transition of global fisheries to sustainable management will secure these benefits for the long term.

5. Only sustainably managed fish stocks can ensure the **viability** of these livelihoods

and, following recovery, generate more employment in the long term. For example, the Ben Tre Clam fishery in Vietnam, after making the transition to sustainability, is now able to support 13,000 households compared to 9,000 in 2007. In light of recent crises, the contribution of fish and fish products to national and global food security has never been of greater importance. Fish is a renewable and healthy food source which currently supplies 1 billion people with their main source of protein.

6. Beyond the direct economic, social and food security benefits to be gained from re-building fisheries, the transition to sustainable management is likely to make marine ecosystems more resilient to external stresses, including those stemming from climate change and pollution. Fisheries are vital components of ecosystems, and healthy eco-systems are keys to the continued productivity of fisheries.

7. Climate change is one of the biggest threats to marine capture fisheries and there is re-search suggesting that those countries most vulnerable to its impacts are often amongst the poorest and most **reliant** on wild fish for food security.

8. Tools for rebuilding global fish stocks—smart economics, the ecosystem approach, ro-bust management

9. Research commissioned by the ISU shows that there are many fisheries around the world that have already **embarked** in these fisheries are varied and context-specific. There is no universal method. However, it is possible to group the available tools un-der three headings: smart economics; an ecosystem approach; and robust manage-ment. Before these instruments can be implemented, there is one **overarching** re-quirement: good governance involving comprehensive stakeholder engagement.

10. Good governance is a **prerequisite** for sustainable and profitable fisheries. Frame-works which uphold good governance are accountable, transparent, responsive, effi-cient and subject to the rule of law.

11. And while top-down frameworks are most effective when they allow for differences at a local level, stakeholder participation at every level, particularly by the fishers themselves, has emerged as one of the most crucial elements of good fisheries gov-ernance.

12. Many of the case studies outlined in this report highlight the importance of co-man-agement models as a way of incorporating stakeholders into decision making proces-ses.

Tools for smart economics

13. The implementation of smart economics has improved the sustainability of many fish-

eries. In some instances this has been achieved through the introduction of limits on the capacity and usage of fishing vessels (effort restriction) and the **allocation** of access rights.

14. As in so many other economic sectors dependent on natural resources, having the right to a portion of these resources creates **incentives** for fishers to maintain the resource in the long term. There are many different ways to establish access rights. They vary from the more advanced individual transferable quota (ITQ) system, implemented with success in countries such as New Zealand and the US; to the territorial user rights (TURF) system.

Tools for an ecosystem approach

15. An ecosystem approach, defined as one that reflects the diverse and dynamic nature of marine ecosystems, is essential to sustainable fisheries management. Many tools exist to help implement the ecosystem approach. These include:

16. Data collection and analysis:
It is difficult to make effective management decisions in the absence of good, usable data. Collaboration between scientists and fishers can be extremely valuable.

17. Precautionary management:
Flexible and dynamic management practices are needed to cope with the uncertainty inherent in marine ecosystems. An example is that of real-time management in the Spencer Gulf **prawn** fishery in Australia that enables changes to fishing activities to be made in just one hour in the event of **undersize** prawn catches.

18. Managing competing users:
In areas where there are competing users, such as the energy, **extractive** and tourist industries, marine **spatial** planning with comprehensive stakeholder engagement is important.

19. Establishing protected areas:
Permanent, temporary or rolling closures of some areas of the marine environment can have benefits for both conservation and fishing activities. One example of this is in the Mediterranean where the fishers of the Prud'hommes de la pêche organization have implemented their own protected area and have noted larger and more abundant fish as a result.

20. Reducing **bycatch**:
The reduction in the catch of non-target species is critical to the sustainability of a fishery and the maintenance of ecosystem health. There are many examples of techniques being developed that aim to reduce bycatch. The "eliminator" trawl is one

such innovation which is able to differentiate between cod and haddock behavior, and thus reduce the catch of non-quota cod.

Tools for robust management

21. Robust management through monitoring and enforcement ensures **compliance** with sustainable fishery goals and regulations. This is necessary for the creation of a level playing field, whereby fishers can operate in the knowledge that the resource is not being overexploited by others.

22. Significant steps are being taken in fisheries across the world to move towards more sustainable management and these fisheries provide grounds for considerable optimism. However, many fisheries are still in a **perilous** state, not least because the tools and interventions highlighted in the case studies are not being deployed, or if they are, at inadequate scale.

23. In turn, this reflects the fact that transition from business as usual to a sustainable state is a complex process which involves tradeoffs and losers as well as winners, especially in the short term. In many cases, the transition requires upfront investment, and leads to temporary decreases in jobs and income, before the economic, social and environmental benefits are realized. It requires strong will and leadership from the fishing industry, government and local communities.

24.   The critical question, therefore, is how can the transition to sustainable fisheries management be more widely enabled and the pace of change accelerated? This report suggests that there are three key enablers of change:

25. Increasing knowledge and awareness of the importance of sustainable fisheries, both by raising the relative importance given to fisheries in international and national discussions and by increasing data collection and collaborative research between scientists and the fishing industry;

26. The provision of significant funding for fisheries in transition so as to finance management changes and tools, and to **mitigate** the impact of short-term reductions in fishing activity and incomes;

27. Greater participation from the private sector along the supply chain, in the form of support for fisheries improvement projects, demand for seafood that is certified sustainable, and taking responsibility for supply chain **traceability** and the sustainability of inputs such as fishmeal and fish oil.

(1,355 words)

➤ *New Words*

scalable [ˈskeɪləb(ə)l]　　*a.* capable of being scaled; possible to scale 可攀登的

As a search company, Google try to develop scalable solutions to problem.

作为一家搜索公司,谷歌致力于开发可升级的问题解决方案。

deplete [dɪˈpliːt]　　*v.* to use up (resources or materials) 耗尽,用尽

That drought has depleted the region's major water supplies.

那场旱灾耗尽了当地主要的水供应资源。

resilient [rɪˈzɪlɪənt]　　*a.* recovering readily from adversity, depression, or the like

有回弹力的;有弹性的;能迅速恢复原状的;迅速恢复活力的

Undoubtedly it gives markets greater depth and continuity, making them more resilient, but it is not without dangers.

毫无疑问,它提高了市场的深度和连续性,增强了市场弹性,但它并非没有危险。

viability [ˌvaɪəˈbɪlətɪ]　　*n.* survival ability; prospective potency 生存能力,发育能力

Any solution has got to be leading to long-term viability.

任何解决方案都应以提供长期生存能力为前提。

reliant [rɪˈlaɪənt]　　*a.* relying on another for support 依赖的

Most companies are now reliant on computer technology.

大多数公司现在依赖计算机技术。

embark [emˈbɑːk]　　*v.* to set out on (an enterprise or subject of study) 使从事,着手

They really don't care how keen you are to embark on a new career.

他们真的不关心你是多么渴望从事一份新的职业。

overarching [ˌəʊvərˈɑːtʃɪŋ]　　*a.* be central or dominant 首要的;支配一切的;包罗万象的

If there is one overarching principle of endurance-building, this is it.

如果提高耐力有一个首要原则的话,那就是它。

prerequisite [priːˈrekwɪzɪt]　　*n.* something that is required in advance 先决条件

In today's Information Age, reading is a prerequisite for success in life.

在当今信息时代,阅读是取得成功的先决条件。

allocation [ˌæləˈkeɪʃ(ə)n]　　*n.* distribution according to a plan 分配,配置

You have to dynamically change the business rules and the human resources allocation, analyze the running process, and survey alerts to adapt to these new requirements.

你必须动态地更改业务规则和人力资源分配,分析正在运行的

流程,并对警报进行核查,以适应这些新的需求。

incentive [ɪnˈsentɪv]　　　　*n.* a positive motivational influence 动机;刺激

If they have a direct incentive to do so they will think about it.

只有他们有了直接动机,他们才会思考这样做的意义。

prawn [prɔːn]　　　　*n.* shrimp-like decapod crustacean having two pairs of pincers 对虾

Draw a dress and a prawn.

画一件服装和一只对虾。

undersize [ˈʌndəˈsaɪz]　　　　*a.* smaller than normal for its kind 不够大的;小于一般尺寸的

Bolts with drilled or undersize heads shall not be used.

不能使用被钻孔或头部尺寸太小的螺栓。

extractive [ɪkˈstræktɪv]　　　　*n.* something extracted or capable of being extracted 提取物;抽出物

Investment in complex projects, mostly in infrastructure and the extractive industries.

对复杂项目进行投资,主要是指对基础设施和采掘工业项目进行投资。

spatial [ˈspeɪʃ(ə)l]　　　　*a.* involving or having the nature of space 空间的;存在于空间的;受空间条件限制的

So, what do we know about spatial distribution?

关于空间分布,我们了解哪些东西?

bycatch [ˈbaɪkætʃ]　　　　*n.* unwanted marine creatures that are caught in the nets while fishing for another species 附带捕捞

Bycatch must not be discarded. Instead it must be landed and recorded as part of that boat's quota.

附带捕捞物不允许丢弃,相反它必须上岸卸载并作为这艘船的配额的一部分被记录下来。

compliance [kəmˈplaɪəns]　　　　*n.* acting according to certain accepted standards 顺从,服从,遵照,按照

It is important to drive in compliance with the traffic laws.

行驶时遵守交通规则是很重要的。

perilous [ˈperɪləs]　　　　*a.* fraught with danger 危险的,冒险的

Often, they venture to some of the most perilous places on earth.

他们经常勇敢地踏入地球上最危险的一些地方。

mitigate [ˈmɪtɪgeɪt]　　　　*v.* to lessen or to try to lessen the seriousness or extent of (使)缓和,(使)减轻

New economic measures help mitigate the effects of the recession.

新经济措施有助于减轻经济衰退的影响。

traceability [ˌtreɪsəˈbɪlɪtɪ]　　　　*n.* the ability to verify the history, location, or application of an item by means of documented recorded identification 可追溯性;

跟踪能力

That way, whatever legislation comes down, whatever audits you have to pass, you have traceability from your software development lifecycle right through to your business process outputs.

通过此种方式,无论采用何种法规,需要通过何种审查,你都拥有从软件开发周期一直到商业过程输出的可追溯性。

➤ *Phrases and Expressions*

seek to 追求;争取;力图
make the transition to 转型,过渡
in light of 根据;鉴于;从……观点
in the absence of 缺乏,不存在;无……时,缺少……时
in compliance with 按照

➤ *Translation*

1. The increased pressure on marine capture fisheries... rising demand for seafood, and a rapid increase in fisheries exploitation—has caused a decline in the productivity of many fisheries. (Para. 2)

   海洋捕捞渔业面临的压力越来越大,来自不断增长的人口、对海产品需求的增加以及渔业的快速增长,这些都导致了许多渔业生产力的下降。

2. As well as delivering less food and income and supporting fewer livelihoods than they could,... to external pressures such as climate change and pollution. (Para. 2)

   除了不能提供更多的粮食和收入,以及支持更多的生计,过度捕捞的渔业也更容易受到气候变化和污染等外部压力的影响。

   【该句中使用了比较句式 less... than... 的结构,在英语中一些比较句式在汉语中不一定含有比较的含义,所以结合上下文的语境,句子想要表述的是渔业过度捕捞的不景气的状况,如果直译则不符合汉语的表达习惯。因此我们可以把此处的比较结构替换为否定+原级结构,即 less= not much,这样使译文内容更贴切,句子通顺。】

3. It seeks to outline the critical importance of wild fish stocks and the benefits that come from their sustainable management. (Para. 3)

   该报告力图概括出野生鱼类资源的重要性及其可持续管理带来的好处。

   【这里主要是指 International Sustainability Unit 国际可持续发展机构实施的研究咨询得出的报告内容,在翻译的时候要将主语补充出来。】

4. It shows how these benefits are already being realized in fisheries around the world through the implementation of a wide range of tried and tested tools for sustainability. (Para. 3)

   报告指出,通过实施多种多样的经过反复测试的可持续性工具,世界各地的渔业是如何实

现这些效益的。

【英语语言习惯是前置性描述,先果后因;汉语则相反,先因后果,层层递进,最后总结。该句中介词短语部分作状语成分,根据汉语表达习惯,在翻译的时候放到句子前面,这样可以使句子意义十分紧凑,结构上整体感更强。】

5. If degraded fisheries are rebuilt and sustainably managed, they can make an even larger contribution. (Para. 4)

   如果退化的渔业得到重建和可持续管理,它们可以做出更大的贡献。

6. Research commissioned by the ISU shows that there are many fisheries around the world that have already embarked in these fisheries are varied and context-specific. (Para. 8)

   ISU 委托授权进行的一项研究表明,世界上有许多渔业已经开始从事这些多样化的、特定背景环境的渔业。

7. And while top-down frameworks are most effective when they allow for differences at a local level, stakeholder participation at every level... has emerged as one of the most crucial elements of good fisheries governance. (Para. 10)

   虽然这种能够考虑地方差异性的、自上而下的管理体系是最有效的,但各级利益相关者的参与,尤其是渔民本身的参与已经成为良好渔业管理的最关键要素之一。

8. Permanent, temporary or rolling closures of some areas of the marine environment can have benefits for both conservation and fishing activities. (Para. 18)

   海洋环境的某些区域永久性的、临时的或定期的封闭对于环境保护和渔业捕捞都是有益处的。

9. Significant steps are being taken in fisheries across the world to move towards more sustainable management and these fisheries provide grounds for considerable optimism. (Para. 21)

   为了朝着更加可持续发展的管理方向迈进,世界各地的渔业都采取了重大的举措,这也让人有充足的理由对这些渔业持乐观态度。

10. In turn, this reflects the fact that transition from business as usual to a sustainable state is a complex process which involves tradeoffs and losers as well as winners, especially in the short term. (Para. 22)

    反过来,这也反映了一个事实,即从一切如常到可持续发展状态的转变是一个复杂的过程,它涉及对各种利益的权衡、盈亏的考虑,尤其是短期内的转变。

    【该句中含有一个同位语从句,在翻译时可以按原文顺序在本位语之后译成一个独立分句形式。可在本位语前增加"这""这样一个"等字样,然后在从句前加冒号或加译"即",起到同位语解释说明的作用。】

 **Exercises**

1. Fill in the blanks with the proper given words, and then translate the sentences into Chinese.

productive	compliance	overexploited	embark
seek to	in light of	transition	stem from
deplete	degrade	robust	overarching
prerequisite	constraints	extractive	

1) A command of information is the necessary _____ to the scientific consideration of any subject.

2) We _____ appropriately address these problems together with the US through equal consultation in a constructive manner.

3) They need to do two things: invest in the _____ capacity of agriculture and improve the operation of food markets.

4) According to the European Commission, 88% of our fish stocks are _____. Of these, 69% are at risk of collapse.

5) At current levels of production, China's proven reserves of coal will be _____ in 50 years, according to BP.

6) This will help to avoid the chance for resource contention that can severely _____ performance, or lead to a record dead lock condition.

7) This task marks the beginning of the _____ from service solution specification into service realization design.

8) In general, the mentor influences what he or she can _____ personal experience, passion, and the constraints of the workplace.

9) Confidence and fear are contradictory states of mind that _____ both our beliefs and attitudes.

10) If we do not pay attention to this _____ goal, we get no results.

11) However, the reality in China is that most firms just meet the first one—in _____ with existing laws, and even this isn't easy for them.

12) First, there are _____ on what they are able to do to mitigate their situation.

13) Although Europe shows signs of recovery, there are plenty of reasons to fear it will not be a _____ one.

14) In addition, the Bank Group will begin requiring disclosure of revenue figures for new major _____ industries projects immediately, and for all projects within two years.

15) I advise you make thorough inquiry about the enterprise before you in it; don't _____ it blind.

2. Translate the following passage into Chinese.

Robust management through monitoring and enforcement ensures compliance with sustainable fishery goals and regulations. This is necessary for the creation of a level playing field, whereby fishers can operate in the knowledge that the resource is not being overexploited by others. Significant steps are being taken in fisheries across the world to move towards more sustainable management and these fisheries provide grounds for considerable optimism. However, many fisheries are still in a perilous state, not least because the tools and interventions highlighted in the case studies are not being deployed, or if they are, at inadequate scale. In turn, this reflects the fact that transition from business as usual to a sustainable state is a complex process which involves tradeoffs and losers as well as winners, especially in the short term. In many cases, the transition requires upfront investment, and leads to temporary decreases in jobs and income, before the economic, social and environmental benefits are realized. It requires strong will and leadership from the fishing industry, government and local communities.

3. Translate the following passage into English.

由于以往的捕捞和养殖多采取无组织、粗放型、甚至是掠夺性的生产方式,这种传统的发展模式已经显露出其负面影响,带来了诸如渔业资源衰竭、生态环境破坏、渔业新技术引入不力等问题,这些问题又直接关系到渔业的生存和持续发展。渔业的可持续发展是我国农业乃至整个国民经济可持续发展的一个重要组成部分。渔业经济可持续发展要求渔业生产的增长不能以牺牲环境为代价,资源的利用与环境的保护应尽可能满足当代人的需求而不牺牲后代人的需求。

Unit 2
Enhancement of Fishery Resources

Adjusting Water Discharge Can Boost Salmon Productivity

1. By adjusting water **discharges** in ways designed to **boost** salmon productivity, officials at a dam in central Washington were able to more than **triple** the numbers of **juvenile** salmon downstream of the dam over a 30-year period, according to a study published in the *Canadian Journal of Fisheries and Aquatic Sciences*.

2. "This is one of the most productive populations of fall Chinook salmon anywhere in the Pacific Northwest," said Ryan Harnish, first author of the paper and a fish ecologist at the Department of Energy's Pacific Northwest National Laboratory.

3. "Much of this productivity can be attributed to the way in which operations of Priest Rapids Dam have been **altered** over the past 30 years. "

4. The major point of the paper is simple: Keeping eggs and young salmon under water, especially at their most vulnerable times, boosts their survival. "The timing of life stages in relation to water temperature has been studied so extensively for this population that we have a good idea of when the eggs will be **hatching**, when the young fish will emerge from the **gravel**, and when the **smolts** will migrate to sea," added Harnish. "This allows for water flows to be planned around these life events, to minimize the risk that the eggs and the young fish below the dams will be **dewatered**. "

5. While dams are known for the physical barriers they impose to migrating salmon, the way that dams hold back and release water for flood control and production of electricity can also have a huge impact. Two of the biggest risks are "dewatering"—when water levels drop so low that young fish or eggs don't stay **immersed** in water—and when water levels fall quickly, leaving young fish **stranded** on the shoreline or **entrapped** in pools of water that can heat up or dry out.

Priest Rapids Dam and the Hanford Reach

6.   Harnish and his team analyzed 35 years' worth of data, from 1975 through 2009, about salmon downstream from the Priest Rapids Dam in an area known as the Hanford Reach. The section of the Columbia River, about 50 miles long and roughly between the communities of Mattawa and Richland, borders the Hanford Site and is one of the longest free-flowing sections of the Columbia River available to salmon. The salmon that spawn in this span of the Columbia—known specifically as upriver bright fall Chinook salmon—are especially **prized** for their size and quality.

7.   In the mid-1970s, several major dewatering events downstream of Priest Rapids Dam resulted in huge losses of juvenile salmon and **spurred** two agreements to boost salmon survival. Harnish's study spanned years before either agreement as well as after the two agreements:

8.   The 1988 Vernita Bar Settlement Agreement (VBSA), designed to prevent dewatering eggs and newly hatched fish, known as alevins, by boosting minimum water levels during **incubation**. The agreement also altered the pattern of discharge during the spawning period to prevent salmon from **depositing** eggs in areas that could not be kept under water during the lower flow conditions of winter. The 2004 Hanford Reach Fall Chinook Protection Program Agreement (HRFCPPA), which maintained constraints set forth by the VBSA and also limited the magnitude of flow fluctuations during the period of juvenile rearing to reduce stranding and entrapment.

Study results

9.   Altogether the team found a 283 percent increase in the freshwater productivity of the Hanford Reach fall Chinook salmon population during the years when the fish protection agreements were in place compared to the period before either agreement was enacted. The study estimated that in the most recent years of the study, more than 52 million juvenile salmon were produced annually, on average, in the Hanford Reach—higher than most people imagined. That compares to about 14 million annually, on average, before either agreement took effect.

10.   "One takeaway message from our study is that the constraints agreed upon 10 and 26 years ago are still working to protect salmon in the Hanford Reach," said Harnish.

11.   The majority of the increase in productivity was due to constraints enacted as part of the VBSA to prevent the dewatering of salmon nests known as redds. The team saw additional benefit from changes designed to limit stranding and entrapment of young fish, which allowed more of them to reach the pre-smolt stage, when they are big enough to migrate downstream.

12.   "This program is an excellent example of how commitment to sound science and adaptive management can help identify the appropriate balance between resource use and protections," said Russell Langshaw, a co-author of the paper and a fisheries scientist at Grant County Public Utility District, which runs the Priest Rapids Dam and which funded the study. "Operations at Priest Rapids Dam are altered dramatically to meet constraints of the HRFCPPA. We're able to provide these protections and meet load demand because of the strong commitment by all operators on the mid-Columbia to act as a coordinated system." In a time when abundance and productivity are declining for many salmon stocks, it's exciting and rewarding to contribute to increases for such an important one," added Langshaw.

13.   The investigators say the results also show how the existence of dams can actually improve salmon survival. For instance, prior to the development of the hydroelectric system, water flow in the Columbia could drop very low in winter. These low flows, when combined with temperatures below freezing, likely resulted in **mortality** of eggs and young salmon still in the gravel. Current dam operations maintain minimum water flows that are more than twice as high as historic levels during winter, which keeps young fish in water more consistently than what might occur naturally.

14.   "Our study shows that dams can be used to produce electricity effectively, and with constraints that have been well thought out, those operations can not only have reduced negative impact, but they can positively impact salmon production," said PNNL ecologist Geoff McMichael, an author of the paper.

15.   In addition to Harnish, Langshaw and McMichael, other authors of the paper include Todd Pearsons of the Grant County Public Utility District; and Rishi Sharma, formerly of the Columbia River Treaty Inter-Tribal Fish Commission and now with the Indian Ocean Tuna Commission.

(1,034 words)

➢ *New Words*

discharge [dɪs'tʃɑːdʒ]

n. when gas, liquid, smoke, etc., is sent out, or the substance that is sent out; the substance that comes out 排放;排出
Reducing the impacts of effluent discharge, improving water quality and responsible use of water are key areas to be considered in aquaculture development.
减少污水排放的影响、改善水质和负责任地使用水资源是水产养殖发展需要考虑的关键问题。

boost [buːst]

v. to increase or improve something and make it more successful 促进;增加;支援
The new resort area has boosted tourism.
新的度假区促进了旅游业的发展。

triple ['trɪp(ə)l]

v. to increase by three times as much 使成三倍,增至三倍
We expect to triple our profits next year.
我们期望明年利润增至三倍。

juvenile ['dʒuːvənaɪl]

a. displaying or suggesting a lack of maturity 幼小的;处于生长期的;青少年的
Juvenile period is one in which the self-consciousness develops rapidly and the outlook on life comes into being gradually.
青少年时期是自我意识迅速发展的时期,是人生观逐步形成的时期。

alter ['ɔːltə]

v. to change, or to make someone or something change (使)改变,更改
In doing translation, one should not alter the meaning of the original to suit one's own taste.
翻译时不应根据自己的好恶改变原文的意思。

hatch [hætʃ]

v. to let the young bird, insect, etc. come out 孵;孵化
The eggs take three days to hatch.
这些蛋要三天时间才能孵化。

gravel ['græv(ə)l]

n. small stones used to make a surface for paths, roads, etc. 碎石;沙砾
All roads near the Odessa highway are more or less fine. But 50 kilometers further they become really terrible, some kind of mixture of asphalt and gravel.
所有临近敖德萨的高速公路都还说得过去。但是 50 千米开外,路就变得非常难走,路面大概是沥青和碎石的

混合物。

smolt [sməult]　　　　　　　*n.* a young salmon at the stage when it migrates from fresh water to the sea 幼鲑;初次由河入海的小鲑鱼;二龄鲑
That is the smolt-to-adult stage of salmon life.
那就是鲑鱼生命中由幼鲑到成鱼的阶段。

dewater [diːˈwɔːtə]　　　　　*v.* to remove moisture 除去……的水分;使脱水
Based on analysis of the characteristics of sludge composition and formation, it is believed that the key point is to dewater and solidify sludge on the spot.
通过对淤泥的组成与产生特性的分析,淤泥的现场脱水固化被认为是关键点。

immerse [ɪˈmɜːs]　　　　　　*v.* to put someone or something deep into a liquid so that they are completely covered 使浸没(于液体中)
Immerse in water for one minute.
浸没在水里 1 分钟。

strand [strænd]　　　　　　　*v.* to make a boat, fish, whale, etc. be left on land and unable to return to the water 使搁浅
The receding tide stranded the whale.
退潮把鲸鱼搁浅在海滩上。

entrap [ɪnˈtræp]　　　　　　　*v.* to trap someone or something, or make it impossible for them to escape from a situation, especially by tricking them 使陷入圈套(困境);诱捕
Discarded fishing lines or nets entrap sea life, amputating fins or strangling them.
废弃的钓鱼线和渔网会困住海洋生物,割伤它们的鳍或扼死它们。

prize [praɪz]　　　　　　　　*v.* to think that someone or something is very important or valuable 珍视
The company's shoes are highly prized by fashion conscious youngsters.
该公司的鞋备受时尚年轻人的青睐。

spur [spɜː]　　　　　　　　　*v.* to encourage someone or make them want to do something 激励,鞭策;促使
It was an article in the local newspaper which finally spurred him into action.
当地报纸上的一篇文章最终促使他行动起来。

incubation [ˌɪŋkjʊˈbeɪʃ(ə)n]　*n.* sitting on eggs so as to hatch them by the warmth of the body 孵化
It is caused during development by high incubation tempera-

tures.

这是在发育阶段由高温孵化造成的。

deposit [dɪ'pɒzɪt]	*v.* to put something down in a particular place 放置(某物);使沉积;存放
	The female deposits her eggs directly into the water.
	雌性把卵直接产在水中。
mortality [mɔː'tælɪtɪ]	*n.* the number of deaths during a particular period of time among a particular type or group of people 死亡率;死亡数
	An investment in safe water and sanitation for homes and schools can be a key factor in reducing child mortality.
	投资于家庭和学校的安全饮用水和卫生设施可以是降低儿童死亡率的一个关键因素。

➤ *Phrases and Expressions*

impose... on 推行,强制实行;施加影响于

hold back 抑制;阻止

heat up 升温;受热;加热

dry out 干透;使变干

set forth 陈述,提出

be in place 在适当的位置

➤ *Translation*

1. By adjusting water discharges in ways designed to boost salmon productivity, officials at a dam in central Washington were able to more than triple the numbers of juvenile salmon... according to a study published in the *Canadian Journal of Fisheries and Aquatic Sciences.* (Para. 1)
   根据一项发表于《加拿大渔业和水产科学》期刊上的研究,为了提高鲑鱼的生产力,通过调整水流排放的方法,华盛顿州中部的一个大坝的管理人员在 30 年间能够让大坝下游的幼年鲑鱼的数量提高 3 倍。

2. "Much of this productivity can be attributed to the way in which operations of Priest Rapids Dam have been altered over the past 30 years." (Para. 3)
   "这种生产力的很大一部分可以归功于在过去 30 年里,普列斯特·拉皮德斯大坝的运作方式的改变。"

3. "The timing of life stages in relation to water temperature has been studied so extensively for this population... when the young fish will emerge from the gravel, and when the smolts will migrate to sea." (Para. 4)
   "对于这种鱼类,与水温有关的生命各个阶段的时间点已经被广泛地研究过了,所以我们很清楚地了解鱼卵什么时候孵化,幼鱼什么时候从沙砾中出现,二龄鲑什么时候迁移入海。"

4. In the mid-1970s, several major dewatering events downstream of Priest Rapids Dam resulted in huge losses of juvenile salmon and spurred two agreements to boost salmon survival. (Para. 7)

在 20 世纪 70 年代中期,普列斯特·拉皮德斯大坝下游的几个大的脱水事件导致了幼年鲑鱼的巨大损失,从而促成了两项提高鲑鱼存活率的协议。

5. The agreement also altered the pattern of discharge during the spawning period to prevent salmon from depositing eggs in areas that could not be kept under water during the lower flow conditions of winter. (Para. 8)

该协议还改变了在鲑鱼产卵期水流量的排放方式,以防止鲑鱼在冬季水流量较小的条件下无法在水中产卵。

6. The study estimated that in the most recent years of the study, more than 52 million juvenile salmon were produced annually, on average, in the Hanford Reach—higher than most people imagined. (Para. 9)

据该项研究估计在最近几年的研究中在汉福德流域幼年鲑鱼的产量平均每年高达 5 200 万,这比大多数人想象的要高。

7. The team saw additional benefit from changes designed to limit stranding and entrapment of young fish, which allowed more of them to reach the pre-smolt stage, when they are big enough to migrate downstream. (Para. 11)

该研究小组发现,为了防止幼鱼搁浅和对捕捞的限制所做的改变还有一些额外的好处,这样的改变可以让更多的幼鱼生长到可以入海前的阶段,这样是为了让幼鱼长到足够大,可以迁徙入海。

【有些定语从句同时具有状语从句的功能,在意义上与主句有状语关系,表原因、结果、目的、让步等关系,在翻译这样的定语从句的时候,要善于从原文中发现句子之间的逻辑关系,然后译成汉语中相应的句子。该句中"when they are big enough to migrate downstream"作定语修饰先行词 the pre-smolt stage,而从句子的逻辑关系来看,该定语从句则表示目的。】

8. This program is an excellent example of how commitment to sound science and adaptive management can help identify the appropriate balance between resource use and protections. (Para. 12)

这个项目是一个很好的例子,说明对正确的科学和适应性管理的保障有助于确定资源使用和资源保护之间的适当平衡。

9. Current dam operations maintain minimum water flows that are more than twice as high as historic levels during winter, which keeps young fish in water more consistently than what might occur naturally. (Para. 13)

目前的大坝运行保持着最低的水流量,水流量是在冬季历史水平的两倍以上,这使得幼鱼在水里的流动比自然发生的情况更加稳定。

【该句中的定语从句"that are more than twice as..."属于限定性定语从句,对主句起着限制作用,按照汉语的表达习惯在翻译时应把从句内容前置,但是由于汉语中断句的使用比英

语更普遍,所以为了句子通顺,可以把定语从句译为与主句并列的句子。】

10. Our study shows that dams can be used to produce electricity effectively, and with constraints that have been well thought out, those operations can not only have reduced negative impact, but they can positively impact salmon production. (Para. 14)

我们的研究表明,大坝可以用来有效地发电,而且经过仔细考虑的约束条件,这样的操作不仅可以减少负面影响,而且还能对鲑鱼的生产产生积极影响。

 **Exercises**

1. Fill in the blanks with the proper given words, and then translate the sentences into Chinese.

triple	productive	fluctuate	adjust
alter	attribute	minimize	emerge
impose	set forth	span	discharge
deposit	magnitude	spur	

1) To generate lift, a bird has merely to tilt its wings, _____ the flow of air below and above them.

2) He reasons that if you can sit in a chair an extra 15 to 20 minutes a day over a year, the savings are probably double or _____ the cost of the chair.

3) As their boss, what can you do to ensure that they are as _____ as possible?

4) We can then translate this information into estimates of the current diabetes and cardiovascular disease that can be _____ to the rise in consumption of these drinks.

5) But for a long time Europeans have _____ their landscapes, including the removal of 90% of wetlands and floodplains, which are crucial parts of river ecosystems, he said.

6) As the next step, whether or not these people have an awakening they _____ from the state in which they have been with important new information about themselves.

7) Therefore, he said, a committee of top officials would work to repeal the report and _____ "the great damage" it has done to the state of Israel.

8) To give a text an author is to _____ a limit on that text, to furnish it with a final signified, to close the writing.

9) I basically know the answer: temperature is a noisy time series, so if you pick and choose your dates over a short time _____ you can usually make whatever case you want.

10) The latest data continue to show that suicide statistics in the Army frequently _____.

11) He _____ from the Canaries in a rubber dinghy, with all food on board locked away.

12) Few of those plants have ever been sanctioned for those emissions, nor were their _____ permits altered to prevent future pollution.

13) Like a real bank, kids can't make a _____ or withdrawal by modifying their passbooks; they have to ask the bank to do it.

14) If that strain was to be released in one event, the quake would top _____ 8. 5, a so-called great earthquake.

15) I welcome this healthy and fair competition because I believe we'll see it will _____ us both to innovate and both will benefit from it.

2. Translate the following passage into Chinese.

The major point of the paper is simple：Keeping eggs and young salmon under water, especially at their most vulnerable times, boosts their survival. "The timing of life stages in relation to water temperature has been studied so extensively for this population that we have a good idea of when the eggs will be hatching, when the young fish will emerge from the gravel, and when the smolts will migrate to sea," added Harnish. "This allows for water flows to be planned around these life events, to minimize the risk that the eggs and the young fish below the dams will be dewatered."

While dams are known for the physical barriers they impose to migrating salmon, the way that dams hold back and release water for flood control and production of electricity can also have a huge impact. Two of the biggest risks are "dewatering"—when water levels drop so low that young fish or eggs don't stay immersed in water—and when water levels fall quickly, leaving young fish stranded on the shoreline or entrapped in pools of water that can heat up or dry out.

3. Translate the following passage into English.

联合国粮食及农业组织的全球鱼类价格指数在五月份创造了一个新的纪录。饮食习惯的改变,尤其是在中国是鱼价持续上涨的主要原因。此外,高昂的油价抬高了捕捞和运输成本,这些成本最后又被转嫁到了餐桌上。然而,鱼的种类有许多,价格也参差不齐。鱼的生产方式有两种:捕捞野生鱼和养殖家鱼。两者的价格趋势不尽相同。相比如三文鱼那样容易养殖的鱼种,金枪鱼这些主要依靠捕捞的鱼种的价格增长就要大得多。全球野生鱼类的捕捞数量在过去二十年中几乎没有变化。在 20 世纪 80 年代末期,就已经达到了 9 000 万吨的极限值。过渡捕捞是一个原因,另一个原因是生产率增长的空间有限,尤其是在消费者对质量有很高要求的情况下。

Unit 3

Fishery Resources Management

Reform of the CFP Required

1. This EU report provides an overview of the EU fisheries policy that benefits from public support but that, at the same time, suffers unwanted side effects.

Introduction

2. The European fishing industry faces **immense** challenges in economical, ecological and social respects. In an effort to support a respective transitioning of the sector, the European Union and its Member States grant **subsidies** to the fishing industry. As over 72 percent of the assessed EU fish stocks are **overfished** and 22 percent fall outside of safe biological limits (European Commission, 2010), one of the main outcomes that is needed as a result of this transition is to bring the fishing capacity in line with the sustainable yield of stocks.

3. Subsides in the fisheries sector may have several negative effects such as creating or maintaining overcapacities or lowering retail prices, thereby increasing additional consumer demand for resources that are already under pressure. From a purely economic point of view, support schemes artificially increase potential revenue. Ultimately, subsidies may create **incentives** for unprofitable fleets to remain in business or to increase their fishing efforts, resulting in **overcapacity** and leading to an overexploitation of the resources (Sumaila and Pauly eds. , 2007). The extent to which fisheries subsidies cause environmentally harmful effects depends on **variables** like the state of the existing management system, the type of fishery, the way in which it **functions** and control and **enforcement** measures as well as the biological status of stocks (Markus, 2010).

4. In the case of the European Union, it remains a challenge to **align** the respective management and control systems in such a way that fisheries subsidies do not cause harm to fish resources. The support schemes under the Common Fisheries Policy (CFP)

have changed in recent years, reflecting the EU Sustainable Development Strategy from May 2001. A variety of subsidies have been **eliminated**, such as the construction of new vessels, and funds have been **redirected** to programs aimed at reducing fleet capacity, but the overall fishing capacity has not been sufficiently reduced to date ("Baltic Sea 2020", 2009).

5. On 22 April 2009, the European Commission published a Green Paper "Reform of the Common Fisheries Policy". The paper illustrates how the **substantial** public financial support for the fisheries sector is often incompatible with, and even **contradictory** to other Common Fisheries Policy (CFP) objectives, particularly the need to reduce overcapacities (European Commission, 2009). In addition, a recent evaluation of the European Fisheries Fund's [Council Regulation (EC) No. 1198/2006] predecessor, the Financial Instrument for Fisheries Guidance (FIFG 2000–2006), demonstrated that Member States generally failed to use environmental or social criteria to guide their decisions on where to **allocate** subsidies. As a consequence, EU fisheries subsidies continued to maintain and even increase fishing overcapacity in a number of fisheries (Cappell et al., 2010).

Overview of fisheries subsidies in the European Union

6. The total value of fisheries subsidies in the European Union is unknown. There are several reasons for this, including that subsidies come from a variety of sources, different definitions of subsidies are applied and serious issues remain with **transparency** aspects and non-compliance with reporting requirements. A recent study estimates that subsidies in the European Union account for about 46 percent of the landed value of fisheries (Sumaila and Pauly eds., 2007).

7. In a number of EU Member States the cost to their national budget of managing and subsidizing fisheries now **surpasses** the economic value of the catches (European Commission, 2008b).

8. The European Fisheries Fund (EFF) is the main structural funding instrument of the European Union in the fisheries sector. The EFF's total budget is about 4.3 billion for the seven-year period from 2007–2013.

9. In addition, the fisheries sector profits from: Support to access the fishing zones of third countries; Community aid to support the controlling and monitoring efforts of the Member States; The EU tax **exemption** on fuel used by vessels. Support received through other structural funds such as the European Regional Development Fund (ERDF) and the European Social Fund (ESF) with the aim of reducing socio-economic disparities; Rescue and restructuring aid such as the emergency aid package to tackle

the fuel crisis, the partial allocation of vessel **decommissioning** aid to firms that switch to smaller, more energy-efficient vessels, temporary reductions in employee contributions to social security payments and emergency aid for the temporary suspension of fishing activities.

Conclusions and recommendations

10. The potential future impacts of subsidies will depend on the state of fish stocks, the type of management regime and on the degree of success at enforcing rules (OECD, 2010). The EU did not succeed in aligning these factors and, consequently, a gap remains between the official sustainability objectives of the CFP and the results actually achieved.

11. Currently, no viable mechanism exists to assess the correspondence between fishing capacity and fishing opportunity. In the absence of reliable information (e. g. between specific species recovery plans and fleet adaptation to the kind required) (Lutchman et al. , 2009), it remains likely that European fisheries subsidies will continue to have harmful effects.

12. As the problem of too many boats chasing too few fish persists, a proactive approach should be taken to guide the funds in the desired direction. Subsidies that are likely to have negative effects in the absence of a reliable regulatory regime should be curbed and phased out.

13. Subsidies that directly contribute to the recovery of stocks and their environment should be supported. Additional efforts in this direction seem necessary as Member States have shown a clear preference in the past for allocating funds to potentially harmful projects, such as towards fleet adaption and modernization rather than nature conservation (Lutchman et al. , 2009).

14. Necessary steps to ensure that public funds contribute to a sustainable fisheries industry include:

15. The improvement of the regulatory regime by: increasing transparency and consistently implementing measures such as the European transparency initiative (European Commission, 2006b; 2007); making the access of Member States to structural funds dependent on the fulfilment of their reporting requirements, especially annual reporting obligations, in an effort to achieve a sustainable balance between fishing capacity and fishing opportunities; strengthening conditionality aspects between receiving subsidies and achieving the objectives of the CFP. Non-compliance with the CFP rules should have an effect on the availability of funds. Operators who have been convicted

of fraud or IUU practices should face financial sanctions and be excluded from receiving future funding; and requiring that any spending under a future financial instrument for the EU fisheries transparently outlines the extent to which the subsidy helps to achieve the objectives of the CFP.

16. Fostering potentially good subsidies, such as: Development of a monitoring and control regime of fisheries; Scientific research for stock assessments; Reduction of impacts on marine habitats and ecosystems; Research and training in the use of environmentally friendly fishing techniques or aquaculture activities; and retraining fishermen for alternative employment opportunities.

17. Curbing or phasing out potentially harmful subsidies, such as: Contributions to operating costs, processing activities or price support; Decreasing aid for individual fishing operations and vessel modification; Payments for fishing access in third country waters which contribute to overfishing as well as political havoc and armed conflict; and phasing out fuel subsidies.

(1,211words)

➢ *New Words*

immense [ɪˈmens]	*a.*	extremely large 巨大的,广大的

But the cost—to the world economy and, above all, to the millions of lives blighted by the absence of jobs—will nonetheless be immense.

但其代价——对整个世界经济,尤其是对那些由于失业而饱受挫折的千百万的民众来说——将仍然是巨大的。

subsidy [ˈsʌbsədɪ]   *n.*   money that is paid by a government or organization to make prices lower, reduce the cost of producing goods, etc. 补贴;津贴;补助金

Government subsidy would not discourage their consumption.

政府补贴不能阻碍消费。

overfish [əuvəˈfɪʃ]   *v.*   to take too many fish from the sea, a river etc., so that the number of fish in it becomes too low 对……进行过度捕捞

As the evidence of continued overfishing shows, however, most of the owners have yet to find, or implement, the right policies.

然而,继续存在的明显过度捕捞的现象表明,大部分沿海国家仍必须寻求或实施适当的政策。

incentive [ɪnˈsentɪv]   *n.*   a positive motivational influence 激励;奖励;诱因

Advocates argue that such incentives would be more effective this time around not only because of design, but also because of timing.

拥护者认为,在这个时候,这种刺激将更加有效,不仅仅因为这种设计,而是因为它的时间点。

overcapacity [əuvəkəˈpæsɪtɪ]   *n.*   the situation in which an industry or factory cannot sell as much as it produces 生产能力过剩

These will only lead to overcapacity in production, increased pressure on environmental resources, and unsustainable economic growth.

这些只会导致产能过剩,环境资源压力加大,以及不可持续的经济增长。

variable [ˈveərɪəbl]   *n.*   a factor that can change in quality, quantity, or size, that you have to take into account in a situation 可变因素

Decisions could be made on the basis of price, delivery dates, after-sales service or any other variable.

决定可以基于价格、送货日期、售后服务或是任何其他可变因素而做出。

function ['fʌŋkʃ(ə)n]　　　*v.* to work in the correct or intended way 起作用,正常运转,运行

Flights in and out of Beijing are functioning normally again.

进出北京的航班又恢复正常了。

enforcement [ɪnˈfɔːsm(ə)nt]　　*n.* the act of enforcing; insuring observance of or obedience to 执行,实施;强制

It provides for regulatory and enforcement tools to go after manipulation, fraud, and other abuses in these markets.

它对操纵市场、欺诈和其他市场上的滥用行为规定了监管和强制手段。

align [əˈlaɪn]　　　　　　*v.* to publicly support a political group, country, or person that you agree with 公开支持;与……结盟;使结盟

Church leaders have aligned themselves with the opposition.

教会领袖和反对派站在同一阵线上。

eliminate [ɪˈlɪmɪneɪt]　　*v.* to completely get rid of something that is unnecessary or unwanted 消除;排除

There is no solution that will totally eliminate the possibility of theft.

没有办法可以完全杜绝盗窃案的发生。

redirect [ˌriːdəˈrekt]　　*v.* to change course or destination 使改变方向;使改变线路

Controls were used to redistribute or redirect resources.

采取了管制措施来重新分配资源或转变资源投放方向。

substantial [səbˈstænʃ(ə)l]　*a.* large in amount or number 大量的

The thesis may at first strike one as extreme, but substantial evidence exists to support it.

这一论文起初给人以偏激的印象,但是有大量的证据支持这一观点。

contradictory [kɒntrəˈdɪkt(ə)rɪ] *a.* conflictive; in disagreement 矛盾的

People find the information in these books contradictory.

人们从这些书上获得的信息是矛盾的。

allocate [ˈæləkeɪt]　　　　*v.* to distribute according to a plan or set apart for a special purpose 分配;指定

It's very important to allocate resources to local communities.

将资源分配给地方社区非常重要。

transparency〔træns'pærənsɪ〕   *n.* the quality of being clear and transparent 透明,透明度
All of your work gives the industry more transparency.
你的工作给了这个行业更多的透明度。

surpass〔sə'pɑːs〕   *v.* to be even better or greater than someone or something else 超越;胜过,优于
With this painting he has surpassed himself.
凭借这幅画,他实现了自身绘画水平的突破。

exemption〔ɪg'zempʃ(ə)n〕   *n.* immunity from an obligation or duty 免除,豁免;免税
You qualify for a tax exemption on the loan.
你有资格获准贷款免税。

decommission〔diːkə'mɪʃ(ə)n〕   *v.* to take (a ship) out of service 使(船)退役;报废(船只)
The airline will decommission 17 of its 102 aircrafts an effort to maintain strong earnings yields. 新加坡航空将让 102 架客机中的 17 架退役,以此维持较高的收益率。

➤ *Phrases and Expressions*

fall outside 超出(能力、理解、考虑、讨论等)范围
in line with 和……一致;按照,依照,符合
aim at 针对;目的在于
be incompatible with 与……不相符的;不符合的
phase out 使逐步淘汰;逐渐停止

➤ *Translation*

1. Subsides in the fisheries sector may have several negative effects such as creating or maintaining overcapacities or lowering retail prices... demand for resources that are already under pressure. (Para. 3)
渔业部门的补贴可能会产生一些负面影响,比如产能过剩持续存在或是零售价格走低,从而增加对已处于压力之下的资源的额外消费需求。

2. In the case of the European Union, it remains a challenge to align the respective management and control systems in such a way that fisheries subsidies do not cause harm to fish resources. (Para. 4)
就欧盟而言,以渔业补贴不会对鱼类资源造成损害的方式,将各自的管理和控制系统统一起来,这仍然是一项挑战。

3. A variety of subsidies have been eliminated, such as the construction of new vessels, and funds have been redirected to programs aimed at reducing fleet capacity, but the overall fishing capacity has not been sufficiently reduced to date. (Para. 4)
已经将各种补贴取消,例如新船的建造,并且已经把这些资金转移到旨在减少船队捕捞能

力的项目上,但是迄今为止,并没有有效地减少总的捕捞量。

【该句中的被动语态的翻译是难点。在英语中被动语态使用范围广泛,而汉语中使用被动语态的范围则相对狭窄得多。所以大多数英语中的被动语态,可以翻译成符合汉语表达习惯的主动语态。尤其是当不知道或不必说出动作的执行者的时候。一般说来,描述发生、存在或消失了什么事物的英语被动句,翻译的时候往往采用无主句,有时可以再动词前面加上"把""使""将""对"等词。】

4. The paper illustrates how the substantial public financial support for the fisheries sector is often incompatible with, and even contradictory to other Common Fisheries Policy (CFP) objectives, particularly the need to reduce overcapacities. (Para. 5)

这篇论文阐述了对渔业部门的大量公共财政支持往往是如何与其他公共渔业政策(CFP)目标相矛盾的,特别是对减少过度捕捞的要求。

5. There are several reasons for this, including that subsidies come from a variety of sources, different definitions of subsidies are applied and serious issues remain with transparency aspects and non-compliance with reporting requirements. (Para. 6)

造成这种情况的原因有很多,包括各种来源的补贴、对补贴的不同定义,决策信息透明度方面的严重问题,以及不符合报告要求的问题。

【由于汉英语言表达习惯上的差异,有时需要改变原文某些词语的词类或句子成分才能有效传达原文的意思,这就是翻译当中的转译法,主要指词类的转译和句子成分的转译。该句中"There are several reasons for this, including..."是主句,including 分词短语做状语。该句所要传达的信息是"有很多原因,包括……",所以接下来 including 后接 that 引导的宾语从句,我们可以试着将从句的动作简化成名词,与主句"原因"相呼应。把 come from 翻译成"来源",把 definitions are applied 翻译成"对……的定义"。】

6. In a number of EU Member States the cost to their national budget of managing and subsidizing fisheries now surpasses the economic value of the catches. (Para. 7)

在许多欧盟成员国中,国家的管理和补贴渔业预算的成本已经超过了捕获量的经济价值。

7. The potential future impacts of subsidies will depend on the state of fish stocks, the type of management regime and on the degree of success... a gap remains between the official sustainability objectives of the CFP and the results actually achieved. (Para. 10)

补贴的潜在影响将取决于鱼类资源的状况、管理体制的类型以及执行规则的成功程度。欧盟没有成功地协调这些因素,因此,在官方的可持续发展目标和实际取得的结果之间依然存在差距。

8. In the absence of reliable information (e. g. between specific species recovery plans and fleet adaptation to the kind required), it remains likely that European fisheries subsidies will continue to have harmful effects. (Para. 11)

在缺乏可靠信息的情况下(例如,特定物种恢复计划和改进所需种类的船队),欧洲渔业补贴仍有可能继续产生有害影响。

【该句中"it remains likely that...",是以 it 作形式主语的主语从句,这种主语从句在翻译

时,可将主句、从句合译为单句,it 一般不需要译出来。另外括号中的内容依然按原句顺序插入在句中即可。】

9. Subsidies that are likely to have negative effects in the absence of a reliable regulatory regime should be curbed and phased out. (Para. 12)

在缺乏可靠的监管机制的情况下,可能会产生负面影响的补贴应该得到遏制,并逐步取消。

10. Additional efforts in this direction seem necessary as Member States have shown a clear preference in the past for allocating funds to potentially harmful projects, such as towards fleet adaption and modernization rather than nature conservation. (Para. 13)

在这个方面的更多的努力似乎是必要的,因为在过去各个成员国已经明显的倾向于将资金分配给可能有害的项目,比如改进船队和更新现代化设备,而不是自然保护。

 **Exercises**

1. Fill in the blanks with the proper given words, and then translate the sentences into Chinese.

substantial	subsidy	immense	overcapacity
align	incentive	variable	redirect
eliminate	aim at	contradictory	exemption
disparity	allocate	surpass	

1) One disease on our agenda causes _____ suffering in large parts of the world, but does its greatest harm in Africa.

2) The great advantage of a _____, indeed, from the standpoint of the public, is that it makes this fact so clear.

3) But _____ will lead inevitably to serious deflationary pressures, starting with the manufacturing and real-estate sectors.

4) In order for the independent director to exercise his rights impartially and equitably, nothing can go without a constraint mechanism in addition to above-mentioned _____ mechanism.

5) Sustainability researchers also face a mountain of _____ in their quest to create accurate predictions about future waste streams.

6) You see an explicit need to _____ health plans more closely with development plans, and you recognize the absolute necessity of collaboration with other sectors.

7) If through strong diplomacy with Russia and our other partners, we can reduce or _____ that threat, it obviously shapes the way at which we look at missile defense.

8) Our goal is to show that these funds can be _____ towards investment in electricity, particularly renewable energy.

9) The most common practice is setting up "one-stop" facilities, which _____ allowing investors to conduct all procedures in one place.

10) This sounds promising, but as far as we know, _____ investigation of the applicable rules has not yet to be done.

11) The assumptions behind our moral reasoning are often _____, and the question of what is right and what is wrong is not always black and white.

12) And do the same for clothing and toy spending—don't add to your budget, but _____ portions of your budget to them.

13) But from a political standpoint, this created a problem: the solar property-tax _____ appeared to benefit a single company.

14) There the people-made pollutants increasingly _____ the ability of air and water to dilute the contaminants to safe levels.

15) To be fair, there is argument about the extent to which government policy was responsible for the spectacular _____ in income growth.

2. Translate the following passage into Chinese.

The European fishing industry faces immense challenges in economical, ecological and social respects. In an effort to support a respective transitioning of the sector, the European Union and its Member States grant subsidies to the fishing industry. As over 72 percent of the assessed EU fish stocks are overfished and 22 percent fall outside of safe biological limits (European Commission, 2010), one of the main outcomes that is needed as a result of this transition is to bring the fishing capacity in line with the sustainable yield of stocks.

Subsides in the fisheries sector may have several negative effects such as creating or maintaining overcapacities or lowering retail prices, thereby increasing additional consumer demand for resources that are already under pressure. From a purely economic point of view, support schemes artificially increase potential revenue. Ultimately, subsidies may create incentives for unprofitable fleets to remain in business or to increase their fishing efforts, resulting in overcapacity and leading to an overexploitation of the resources. The extent to which fisheries subsidies cause environmentally harmful effects depends on variables like the state of the existing management system, the type of fishery, the way in which it functions and control and enforcement measures as well as the biological status of stocks.

3. Translate the following passage into English.

各国均采取了诸多措施以加强对渔业资源的养护与管理,同时各国间在渔业管理与养护中的合作也不断增强,并成为当前国际渔业经济的发展趋势之一。传统海洋渔业开始由自主经营向联合开发转变,这种合作主要是通过建立全球性的、次区域的或区域的合作管理机制来实现。

各国对公海捕鱼进行了严格限制,因而资源保护性的国际渔业开发利用合作越来越成为发展远洋渔业的必要途径。目前,许多学者开始关注对渔业资源保护性开发利用的国际合作问题。

Chapter 3

Marine Products and Processing

Unit 1

Safety of Fishery Products and Causes

Fishery Products Hazards

1. Fish and fishery products **constitute** an important food **component** for a large section of world population, more so in developing countries, where fish forms a cheap source of protein. In the last two decades there has been an increase in awareness about the nutritional and health benefits of fish consumption. The low fat content of some fish and the presence of poly unsaturated fatty acids in red meat fishes which are known to reduce the risks of coronary heart diseases, have increased the dietary and health significance of seafood consumption (Din et al. , 2004). In USA, seafood consumption has increased from an average of 4. 5 kg per person in 1960 to about 7. 0 kg per person in 2002 (NOAA, 2003). Total fishery production in 2001 was reported to be 130. 2 million tons, of which 37. 9 (30%) million tons was from **aquaculture**, 90% of which came from Asian countries.

2. On the other hand, seafood is also known to have been responsible for a significant percentage of food-borne diseases. With increased fish consumption, there is also an increase in the number of food-borne illness associated with fishery products. In the United States, 10% – 19% of food-borne illness involved seafood as a vehicle and between 1993 and 1997, 6. 8% of the food-borne illnesses involved consumption of fish and shellfish (FAO, 2004). In Japan, where seafood consumption is high and raw fish consumption is popular, 70% of food-borne illness is attributed to seafood. These data highlight the importance of seafood as vehicles for human illness.

3. Seafood often harbour infectious agents that are present naturally in the aquatic environment or introduced through human activities. The illness may be due to the infection caused by the microorganism or due to the intoxication by products of **microorganisms**. **Viruses**, bacteria and parasites are the agents of human disease associated

with seafoods. The products of microorganism such as bacterial toxins, algal toxins or products of bacterial **metabolism** such as **histamine** can cause **intoxication** when fish or shellfish harbouring these products are consumed. In the case of live finfish, microorganisms may be associated with gill, gut and skin. Filter feeding shellfish like **mollusks** concentrate different types of microorganisms present in the environment around them. Some of the microorganisms of human the genus Vibrio, but others like Salmonella, E. coli and many viruses are introduced into the aquatic environment through human activities such as **contamination** by domestic sewage. The type of microorganism associated with seafood may vary depending upon whether it is fresh or processed. Most of the outbreaks of illnesses occur in countries where seafood is eaten raw or is inadequately cooked. For example, in Japan where seafood is eaten raw, 70% of food-borne human illness is seafood associated (Scoging, 2003).

4.   Viruses constitute an important cause of seafood-borne diseases. The intensity of human illness is known only from developed countries. For example in the USA, it is estimated that 80% of food-borne illness are due to **enteric** viruses (Sair, et al., 2002). The major limitation in diagnosing viral infections is the nonavailability of techniques or facilities for detecting or identifying viral **pathogens**. The recent advances in molecular biology has helped to develop or improve the detection techniques for viruses but these are still not used commonly in developing countries.

5.   Viruses being obligate **intracellular parasites** cannot multiply in seafood, but molluscan shellfish concentrate these infectious agents in their tissues. This is the reason why molluscan shellfish are associated with most of the well-recorded outbreaks. Nevertheless, very low numbers (1–100) of viruses are known to be capable of causing infections in humans making these organisms all the more important as agents of human illness. The transmission is through fecal-oral route with infected individuals becoming source of viruses to the aquatic environment.

6.   The viruses causing human infections are classified into two groups, viral gastroenteritis and viral hepatitis (Caul, 2000). The illnesses due to viruses are mainly prevalent in countries where eating raw shellfish is a dietary habit. Though there are more than 110 different viruses known to be excreted in human feces, collectively called the "enteric viruses" (Goyal, 1984). Hepatitis A virus and Norwalk viruses are most commonly involved in seafood related illnesses followed by **rotaviruses** and astroviruses which cause infections in children.

7. Hepatitis A virus (HAV) is a single stranded RNA virus that is classified in the genus Hepatovirus in the picornaviridae family and constitutes the most important group of viral pathogen in seafood. The virus has worldwide distribution and several outbreaks of human illness associated with oyster or clam consumption has been reported from countries such as Italy, USA, Austria, China and Sweden (Butt, et al., 2004a). A major outbreak of hepatitis A in Shanghai, China involved 300,000 cases linked to consumption of shellfish from a sewage polluted area and this constitutes the largest foodborne disease outbreak ever reported. This virus is associated with gastrointestinal tracts of humans, and the infection is transmitted by faecal-oral route. This virus **is** more resistant to chlorination than other enteric viruses. The infections by this viral group can lead to severe hepatitis with serious liver damage. Jaundice, as the disease is well known, is the clinical symptom followed by headache and vomiting. The incubation period is about 4 – 6 weeks. Shellfish such as oysters and clams accumulate HAV in their tissues and are most commonly associated with infections. Filter feeding shellfish can concentrate the viruses several fold in their tissue and the infectious dose is low (10-100 viral particles). Consequently shellfish harvested from areas negative for coliform counts may transmit the diseases (Butt, et al., 2004a). HAV are known to survive for extended periods of time in seawater (Croci, et al., 1999). Infected food handlers can also be important sources of contamination for seafood.

(1,058 words)

➤ *New Words*

constitute [ˈkɒnstɪtjuːt]　　　　　*v.* to form or compose 组成

Listening, speaking, reading, and then writing constitute the fundamental order in language learning.

听、说、读、写是组成语言学习的基本顺序。

component [kəmˈpəʊnənt]　　　　*n.* one of several parts of which something is made 组成部分;成分;部件

Enriched uranium is a key component of a nuclear weapon.

浓缩铀是核武器的一个关键组成部分。

aquaculture [ˈækwəkʌltʃə]　　　　*n.* rearing aquatic animals or cultivating aquatic plants for food 水产养殖

Another is a shortage of land and freshwater for use in aquaculture.

另一个障碍是用于水产养殖的土地和淡水资源不足。

microorganism [maɪkrəʊˈɔːgənɪzəm]　*n.* any organism of microscopic size 微生物

The electron beams work by interfering with microorganism DNA.

电子束通过干扰微生物 DNA 从而发挥作用。

virus [ˈvaɪrəs]　　　　　　　　*n.* (virology) ultramicroscopic infectious agent that replicates itself only within cells of living hosts; many are pathogenic; a piece of nucleic acid (DNA or RNA) wrapped in a thin coat of protein 病毒;恶毒;毒害

The virus has infected the operating system of his computer.

病毒已感染了他的计算机的操作系统。

metabolism [məˈtæbəlɪzəm]　　　*n.* the chemical processes in living things that change food, etc. into energy and materials for growth 新陈代谢

You know we have to change our eating habits because our metabolism has slowed down.

你知道因为我们的新陈代谢变慢了所以我们必须改变饮食习惯。

histamine [ˈhɪstəmiːn]　　　　　*n.* amine formed from histidine that stimulates gastric secretions and dilates blood vessels; released by the human immune system during allergic reactions 组胺

Your immune system spots pathogens in your lungs and releases histamine.

你的免疫系统发现你的肺部有病原体后释放组胺。

intoxication [ɪnˌtɒksɪˈkeɪʃ(ə)n]   *n.* the physiological state produced by a poison or other toxic substance 中毒

In the case of water intoxication, hyponatremia extreme conditions can ultimately lead to coma and death.

在水中毒的情况下,极端的低钠血最终会导致昏迷和死亡。

mollusk [ˈmɒləsk]   *n.* invertebrate having a soft unsegmented body usually enclosed in a shell (美)软体动物

Measuring more than 16 inches (406 millimeters), this giant mollusk is the biggest ever found in its subclass—caudofoveata.

超过 16 英寸(406 毫米)的长度,这种大型的软体动物成为尾腔亚纲中迄今为止发现的最大的物种。

contamination [kənˌtæmɪˈneɪʃ(ə)n]   *n.* the state of being contaminated 污染,玷污;污染物

Investigations traced the source of all Ebola Reston outbreaks to one export facility near Manila in the Philippines, but the mode of contamination of this facility was not determined.

调查追踪到所有埃博拉莱斯顿亚型的暴发都源于菲律宾马尼拉附近的一个出口设施,但未能确定这一设施的污染方式。

enteric [enˈterɪk]   *a.* of or relating to the enteron 肠的;肠溶的

At the same time, humans and other higher-end animals kept their enteric nervous system.

同时,人类和其他高级动物都保存了肠神经系统。

pathogen [ˈpæθədʒ(ə)n]   *n.* any disease-producing agent (especially a virus or bacterium or other microorganism) 病原体;病菌

The reservoir of this pathogen appears to be mainly cattle and other ruminants such as camels.

这一病菌的宿主看来主要是家畜和其他反刍动物,例如骆驼。

intracellular [ˌɪntrəˈseljʊlə]   *a.* located or occurring within a cell or cells 细胞内的

The pool expanded a little bit if we ignored one intracellular parasite; it expanded even more when we looked at core sets of genes of 310 or so.

如果忽略掉一个细胞内寄生物的话,这个基因库还能扩展一些;当我们审视核心的一组 310 个基因时,基因库扩大更多。

parasite ['pærəsaɪt]	*n.*	an animal or plant that lives in or on a host (another animal or plant); it obtains nourishment from the host without benefiting or killing the host 寄生虫

They found that one compound killed the parasite at every level of development—from larva to adult.

他们发现,有一种化合物在寄生虫从幼虫到成虫的各个阶段,都能将该寄生虫杀死。

rotavirus [ˌrəutə'vaɪrəs]	*n.*	the reovirus causing infant enteritis 轮状病毒(一种导致婴儿或新生畜胃肠炎的病毒)

Of the children who had the vaccine, 2 in every 100 had at least one severe episode of diarrhoea and vomiting caused by rotavirus during the year.

在接种疫苗的儿童中,每 100 人中有 2 人在一年以内可能有一个是由轮状病毒引起严重腹泻和呕吐的疾病。

➤ *Phrases and Expressions*

attribute to 把……归因于
associated with 与……有关系;与……相联系
be classified into 分类为……
be resistant to 抵抗……的;对……有反抗作用的
survive for 争取,奋斗

➤ *Translation*

1. The low fat content of some fish and the presence of poly unsaturated fatty acids in red meat fishes which are known to reduce the risks of coronary heart diseases, have increased the dietary and health significance of seafood consumption. (Para. 1)
   一些鱼的低脂肪含量以及红肉鱼类中的多不饱和脂肪酸已知可以降低患冠心病的风险,因此增加了食用海鲜在饮食和健康方面的重要性。
   【在进行英译汉时,为了使译文流畅和更符合汉语叙事论理的习惯,在将清英语长句的结构、弄懂英语原意的基础上,可以对句子进行重新组合。】

2. Total fishery production in 2001 was reported to be 130.2 million tons, of which 37.9 (30%) million tons was from aquaculture, 90% of which came from Asian countries. (Para. 1)
   2001 年渔业总产量是 13 020 万吨,其中 3790 (30%)万吨来自水产养殖,其中 90% 来自亚洲国家。

3. The illness may be due to the infection caused by the microorganism or due to the intoxication by products of microorganisms. (Para. 3)

这种疾病可能是由于微生物引起的感染,或者是由于微生物的产品中毒引起的。

【过去分词短语用作定语时,一般置于其所修饰的名词之后,其意义相当于一个定语从句,但较从句简洁,多用于书面语中。在进行英译汉时,可以把过去分词短语翻译成前置定语。】

4. The products of microorganism such as bacterial toxins, algal toxins or products of bacterial metabolism such as histamine can cause intoxication when fish or shellfish harbouring these products are consumed. (Para. 3)

细菌毒素、藻毒素或细菌代谢产物如组胺等微生物的产物,在食用含有这些产物的鱼类或贝类时,会引起中毒。

【拆句法是把一个长而复杂的句子拆译成若干个较短、较简单的句子,通常用于英译汉。本句子就是把一个长的句子拆分翻译,符合汉语习惯。】

5. The recent advances in molecular biology has helped to develop or improve the detection techniques for viruses but these are still not used commonly in developing countries. (Para. 4)

分子生物学的最新进展有助于开发或改进病毒的检测技术,但这些技术在发展中国家仍未得到广泛应用。

6. Nevertheless, very low numbers (1—100) of viruses are known to be capable of causing infections in humans making these organisms all the more important as agents of human illness. (Para. 5)

然而,很低的数量(1-100)的病毒被认为能够引起人类感染,使得这些生物作为人类疾病的媒介而变得更加重要。

【英语表达中经常出现长句,而汉语的表达习惯则是短句居多,在翻译的时候应注意语言习惯的不同,恰当地把英语的长句采用顺译或者拆译的方法翻译成几个短句。】

7. The viruses causing human infections are classified into two groups, viral gastroenteritis and viral hepatitis. (Para. 6)

导致人类感染的病毒分为两类:病毒性肠胃炎和病毒性肝炎。

8. Hepatitis A virus and Norwalk viruses are most commonly involved in seafood related illnesses followed by rotaviruses and astroviruses which cause infections in children. (Para. 6)

甲型肝炎病毒和诺如病毒(又称诺瓦克病毒)最常见于与海鲜有关的疾病,其次是轮状病毒和星状病毒,它们会导致儿童感染。

【英译汉时,被动语态的翻译常常是经常遇到的,"被"字通常由汉语的其他替换词替换或者汉语可以省略表达被动的形式标志"被"字。】

9. The virus has worldwide distribution and several outbreaks of human illness associated with oyster or clam consumption has been reported from countries such as Italy, USA, Austria, China and Sweden. (Para. 7)

意大利、美国、奥地利、中国和瑞典等国家已报告了该病毒在全球范围内的分布,以及与牡蛎或蛤蜊消费有关的几种人类疾病的暴发。

【由于英汉语言表达的习惯差异,通常汉语的状语位于句首,而英语的状语可前可后,以后为多。汉语是一种意合的语言,在英译汉中要适当地把状语提到句首。】

10. The incubation period is about 4-6 weeks. Shellfish such as oysters and clams accumulate HAV in their tissues and are most commonly associated with infections. (Para. 7)

潜伏期大约为4-6周。贝类如牡蛎和蛤蜊在它们的组织中积累了很多HAV,它们通常与感染有关。

 **Exercises**

1. Fill in the blanks with the proper given words, and then translate the sentences into Chinese.

be responsible for	highlight	harbour	contamination
detect	identify	obligate	transmission
excrete	distribution	sewage	be transmitted by
accumulate	clinical	survive for	

1) Of course we must _____ that the system can hardly be inaccurate.

2) My answer is this: the separation allows us to better understand what is what, why something is needed, who has to do it and _____ the governance and for the delivery of results.

3) There is no evidence of _____ in our lab, and we have controlled for that all along.

4) As for people who love us but whom we do not love, we may be indifferent, or at least would not _____ such a deep overall concern.

5) Basically, one node will heartbeat to the disk and the other will _____ it.

6) Now the government is responding to the public outcry with a draft revision of ethical standards for teachers that would _____ them to protect students.

7) He always _____ other's happiness with his own.

8) When the _____ is good, foreign stations can be heard.

9) We are part of the trajectory of living tissue: our flesh must breathe, metabolize, mate, _____, and eventually die.

10) The principle of _____ in socialist society is "from each according to his ability" and "to each according to his work".

11) If confirmed, BU will be the only known mycobacterial disease to _____ insects.

12) I know more about the north Gaza _____ treatment works than you can possibly imagine.

13) He will spend money too liberally to _____ much wealth.

14）A conservative management style that can balance risk—because companies don't _____ the long term unless they take risks, with safety.

15）On an individual patient basis, initial treatment decisions should be based on _____ assessment and knowledge about the presence of the virus in the community.

2. Translate the following passage into Chinese.

On the other hand, seafood is also known to have been responsible for a significant percentage of food borne diseases. With increased fish consumption, there is also an increase in the number of food-borne illness associated with fishery products. In the United States, 10%–19% of food borne illness involved seafood as a vehicle and between 1993 and 1997, 6.8% of the food-borne illnesses involved consumption of fish and shellfish (FAO, 2004). In Japan, where seafood consumption is high and raw fish consumption is popular, 70% of food-borne illness is attributed to seafood. These data highlight the importance of seafood as vehicles for human illness.

3. Translate the following passage into English.

随着我国水产养殖集约化程度不断提高，水产品质量安全问题日益受到关注。它是"食物安全"的重要组成部分，关系到食物安全、消费者健康和水产品市场竞争力，直接制约我国渔业的可持续发展。本文从江苏省水产品质量安全建设的实践出发，针对目前我国水产品质量安全问题，就渔药残留、农药残留、激素残留、微生物污染、重金属含量等有毒有害物质超标等现象，分析了鱼苗、鱼种、渔药、饲料和环境污染等几个环节的影响因素，围绕水产品质量安全监管提出了对策及建议。

Unit 2
Strategies for Prevention

Fishery Products Hazards

1. Finfish constitute an important part of the modern food industry. We consume more and more products coming from the four corners of the globe and fish, in particular, may be caught thousands of kilometers from their place of consumption. In the early 1990s, the crisis of the fishing industry resulted in both economic and human trage-dies. Drawing conclusions from the crisis the fishing fleets experienced at that time, some groups of producers started to think of better **enhancement** of their production. Since then, they have tried to adapt their production to market demand. This has led them to set up quality **initiatives** (Brittany Quality Sea food, Brittany headland an-glers, etc.), thus promoting their savoir-faire, the freshness certification, or the im-provements in the quality of their products.

2. In the interest of public health, it is vital that both domestically-processed and impor-ted seafood are safe, wholesome, and properly labeled. FDA has had a strong regula-tory program in place since the mid-1990s to ensure the safety of domestic and impor-ted seafood. In fact, the hazard analysis and risk-based preventive controls framework of FDA's seafood-safety program is a basis for the preventive controls requirements for other FDA-regulated foods called for in the FDA Food Safety Modernization Act (FSMA), enacted in 2011. For this reason, FSMA specifically exempts seafood from some of its requirements. However, FSMA also provides the Agency with a number of new authorities that will help improve the safety of domestic and imported FDA-regu-lated foods, including seafood.

3. Because fish are cold-blooded and live in **aquatic** environments, fish and fishery prod-ucts pose unique food safety challenges, which are quite different from those posed by land animals. FDA has developed extensive expertise in these areas over decades of regulating this commodity. Experts in FDA's Center for Food Safety and Applied Nu-trition (CFSAN) are responsible for evaluating the hazard to public health presented by chemical, including toxins, and microbiological contaminants in fish and fishery

products. FDA operates the Gulf Coast Seafood Laboratory in Alabama, which specializes in seafood microbiological, chemical, and toxins research. In addition, seafood research is conducted at CFSAN's research laboratory in College Park, Maryland. FDA, in **collaboration** with the National Oceanic and Atmospheric Administration at the Department of Commerce, also represents the United States at the Codex Alimentarius Commission's Committee on Fish and Fishery Products, the international food safety standard-setting body for this commodity.

4. The Agency has a variety of tools to ensure compliance with seafood safety requirements, including inspections of domestic and foreign processing facilities, examination and sampling of domestic seafood and seafood offered for import into the United States, domestic **surveillance** sampling of imported products, inspections of seafood importers, evaluations of filers of seafood products offered for import, and foreign country program assessments. FDA works closely with our foreign, Federal, state, local, and Tribal partners to share relevant information and ensure that products in US commerce meet applicable FDA requirements.

5. Seafood is one of the most highly-traded commodities in the world. The Agency recognizes that success in protecting the American public depends increasingly on our ability to reach beyond U. S. borders and engage with its government regulatory **counterparts** in other nations, as well as with industry and regional and international organizations, to encourage the implementation of science-based standards to ensure the safety of products before they reach our country.

6. Following the outbreak, the concept of **traceability** of food products has become a matter of special interest to policy makers and scientists. Traceability of finfish products. Specification on the information to be recorded in captured finfish distribution chains, specifies the information to be recorded in marine-captured finfish supply chains in order to establish traceability.

7. Food trade is one of the largest global businesses today and traceability throughout the food supply chains has gained considerable importance over the last few years (Carriquiry and Babcock 2007; Jansen-Vullers et al, 2003; Madec et al, 2001; Mckean, 2001; Thakur and Hurburgh, 2009). Consumers all over the world have experienced various food safety and health issues. In addition to this, the consumer demand for high quality food and feed product, non-GMO (genetically modified organisms) foods and other **specialty** products such as organic food has grown in the past years. These factors have led to a growing interest in developing stems for food supply chain traceability. A number of food safety and traceability laws exist in different countries. The European Union law describes traceability as an ability to track any food.

8. It specifies how traded fishery products are to be identified, and the information to be generated and held on those products by each of the food businesses that physically trade them through the **distribution** chains. The standard deals with the distribution for human consumption of marine-captured finfish and their products, from catch through to retailers or caterers.

9. The ISO definition of traceability concerns the ability to trace the history, **application** and location of that which is under consideration, and for products this can include the origin of materials and parts, the processing history and the distribution and location of the product after delivery. Traceability includes not only the **principal** requirement to be able to physically trace products through the distribution chain, from origin to destination, but also to be able to provide information on what they are made of and what has happened to them. These further aspects of traceability are important in relation to food safety, quality and labelling.

10. Regarding recent food crisis, **legislation** often requires traceability to facilitate the recall of products or to prevent them from reaching the consumer. Mr. Rolf Duus the secretary of the Working Group that developed the standard comments, "In the last few years, there has been an increased interest in traceability and the responsibility for the supply of safe, healthy and **nutritious** food is shared between all actors involved in the production, processing, marketing and consumption of fish and seafood. ISO provides a generic basis for traceability and will help to guarantee the health protection of consumers and ensure fair practices in food trade of finfish products." Potential users of the new standard include: Fishing vessels, Vessel-landing businesses and auction markets. Processors, transporters and storers, traders and wholesalers and retailers and caterers.

11. Oversight of the safety of the U. S. food supply continues to be a top priority for FDA. The Agency has a strong regulatory program in place for seafood products. FDA will continue to work with our domestic and international partners to ensure the safety of both domestic and imported seafood.

(1,083 words)

➤ *New Words*

enhancement [ɪnˈhɑːnsm(ə)nt]　　*n.* an improvement that makes something more agreeable 增加;放大

For the last enhancement, you will introduce another function in the same manner as above.

对于最后一个增强,您将按照与上面相同的方式引入另一个功能。

initiative [ɪˈnɪʃətɪv]　　*n.* readiness to embark on bold new ventures 主动权;首创精神

He showed little initiative, handling all matters strictly by the book.

他缺乏首创精神,墨守成规地办一切事情。

aquatic [əˈkwætɪk]　　*a.* relating to or consisting of or being in water 水生的;水栖的

This study, for the first time, brings all our knowledge together under one global model of water security and aquatic biodiversity loss.

这项研究第一次在水安全和水生生物多样性的全球模型中集中了我们所有的知识。

collaboration [kəˌlæbəˈreɪʃ(ə)n]　　*n.* act of working jointly 合作;勾结

But the value is in the collaboration, not in open source itself.

但是这种价值来源于合作,并非开源本身。

surveillance [sɜːˈveɪləns]　　*n.* the act of carefully watching a person suspected of a crime or a place where a crime may be committed 监视,看守

So I think surveillance for these kinds of things should be increased.

所以我认为对这类事情的监视应该有所提升。

counterpart [ˈkaʊntəpɑːt]　　*n.* a person or thing having the same function or characteristics as another 职位(或作用)相当的人;对应的事物

As soon as he heard what was afoot, he telephoned his German and Italian counterparts to protest.

他一听到在进行中的事,马上就给德国和意大利相应人员打电话抗议。

traceability [ˌtreɪsəˈbɪlətɪ]　　*n.* the ability to trace something 可追溯性;跟踪能力;可描绘

But after we understand the benefits, how do we go about implementing traceability?

但当我们了解了好处之后,我们如何着手实现可追溯性呢?

specialty ['speʃ(ə)ltɪ]   *n.* a subject or job that one knows a lot about or has a lot of experience of 专业,专长

Unless your physics specialty is deeply theoretical or has to do with astrophysics, you can usually find an industry job that is related to your work.

除非你的物理学专业极端理论化,或者与天体物理有关,否则你通常是可以在工业界找到一个相关的职位的。

distribution [dɪstrɪ'bjuːʃ(ə)n]   *n.* the act of sharing things among a large group of people in a planned way (statistics) an arrangement of values of a variable showing their observed or theoretical frequency of occurrence (分布;分配)

The principle of distribution in socialist society is "from each according to his ability" and "to each according to his work".

社会主义社会的分配原则是"各尽所能,按劳分配"。

application [ˌæplɪ'keɪʃ(ə)n]   *n.* the use of a rule of piece of knowledge in a particular situation 应用

The wide application of electronic computers in science and technology will free man from the labour of complicated measurement and computation.

电子计算机在科学技术方面广泛的应用将使人们从复杂的计量和计算中摆脱出来。

principal ['prɪnsəp(ə)l]   *a.* most important element 主要的

Teaching is her principal source of income.

教书是她的主要收入来源。

legislation [ledʒɪs'leɪʃ(ə)n]   *n.* law enacted by a legislative body 立法;立法过程

The new government dismantled their predecessors' legislation.

新政府废除了前任政府的立法。

nutritious [njuː'trɪʃəs]   *a.* of or providing nourishment 有营养的,滋养的

This will involve improving access to nutritious food, to micronutrient supplementation and in many places to preventing infections as well.

这将包括改善营养食品和微量营养素补充品的获得,在许多地方还包括预防感染。

➤ *Phrases and Expressions*

adapt to. . . 使自己适应于……

in the interest of 为了；为了……的利益

a basis for 奠定基础

in relation to 关于

in place 适当，适当的；在适当的地方，在恰当的位置

➤ *Translation*

1. Drawing conclusions from the crisis the fishing fleets experienced at that time. . . producers started to think of better enhancement of their production.（Para. 1）

   从当时的捕鱼船队经历的危机中得出经验，一些生产商开始考虑更好地提高他们的产量。

2. In fact, the hazard analysis and risk-based preventive controls framework. . . is a basis for the preventive controls requirements for other FDA-regulated foods called for in the FDA Food Safety Modernization Act（FSMA）.（Para. 2）

   实际上，FDA 海产品安全计划的危害分析和基于风险的预防性控制框架，是 FDA 食品安全现代化法（FSMA）要求的其他 FDA 管理食品的预防性控制要求的基础。

3. Because fish are cold-blooded and live in aquatic environments. . . which are quite different from those posed by land animals.（Para. 3）

   因为鱼是冷血动物，生活在水生环境，鱼类和渔业产品对食品安全带来独特的威胁，这些威胁与那些陆地动物带来的威胁有很大的不同。

4. The Agency has a variety of tools to ensure compliance with seafood safety requirements. . . examination and sampling of domestic seafood and seafood offered for import into the United States.（Para. 4）

   食品药品管理机构有各种各样的工具来确保符合海产品安全要求，包括对国内和国外加工设施的检查，对国内海产品和进口到美国的海产品的检验和取样。

   【结构重组翻译：有很多翻译的句子，其语法句式结构等完全没有一点中文习惯，这个时候就得分析好句子结构，然后按照自己的表述习惯，将句子结构打乱，再按符合翻译的要求进行句子结构重组，组合出新的句子以准确地表达出所要表达的含义。】

5. The Agency recognizes that success in protecting the American public depends. . . engage with its government regulatory counterparts in other nations.（Para. 5）

   该机构认识到，保护美国公众的成功越来越依赖于其走出境外，并与其他国家的政府监管机构进行接触的能力。

6. Specification on the information to be recorded in captured finfish distribution chains. . . finfish supply chains in order to establish traceability.（Para. 6）

   捕捞有鳍鱼分销链中记录的信息规定了在海洋捕捞的有鳍鱼供应链中要记录的信息，以建

立可追溯性。

【对于一些句子很长但又结构很复杂的句子,一次性将它们直接翻译出来有困难,这时我们可以考虑将整句进行合理拆分,将其分为若干个相对独立的小语句,翻译时候对这些小而简单的句子依次翻译,之后再以合适的连接词将它们串接起来,就能完成句子的翻译工作。】

7. These factors have led to a growing interest in developing stems for food supply chain traceability. (Para. 7)

   这些因素导致了人们对开发食品供应链可追溯性的兴趣越来越大。

8. The standard deals with the distribution for human consumption of marine-captured finfish and their products, from catch through to retailers or caterers. (Para. 8)

   该标准涉及可食用的从海洋捕捞的有鳍鱼及其产品的经销,包括从捕获到零售商或食品供应商这一供应链。

9. Traceability includes not only the principal requirement to be able to physically trace products through the distribution chain… but also to be able to provide information on what they are made of and what has happened to them. (Para. 9)

   可追溯性不仅包括能够通过分销链从原产地到目的地追踪产品的基本要求,而且还包括能够提供关于产品的构成和加工情况的信息。

10. The Agency has a strong regulatory program in place for seafood products. (Para. 11)

    该机构对海产品有强有力的监管程序。

 **Exercises**

1. Fill in the blanks with the proper given words, and then translate the sentences into Chinese.

adapt to	enhancement	consumption	initiative
collaboration	aquatic	compliance	counterpart
surveillance	distribution	traceability	specialty
principal	application	legislation	

1) Water _____ decreased during the winter.

2) Both sites will also need to view, create, and modify change requests, _____ requests, and defect reports.

3) I put it all down to his hard work and _____.

4) This unique creature is part of an imperiled community of _____ organisms in the Gulf of Mexico, where poor water quality and habitat loss have weakened the ecosystem.

5) There is great opportunity for _____, but this is yet to be grasped.

6) To do that with confidence, you need to measure your _____ to them.

7) The police kept the criminal under strict _____.

8) Each of them has a _____ column in the database tables.

9) After creating the requirement type, we should enter all scenarios and set _____ from use cases to these scenarios, as shown in Figure 9.

10) You should get to know recruiters and search firms in your area of _____.

11) The present _____ policy demotivated people.

12) Your _____ to join the club was honoured.

13) Today the _____ tools for prospecting the brain are electrical.

14) There should be nothing controversial about this piece of _____.

15) How do you _____ that, especially with the lack of resources there?

2. Translate the following passage into Chinese.

Finfish constitute an important part of the modern food industry. We consume more and more products coming from the four corners of the globe and fish, in particular, may be caught thousands of kilometers from their place of consumption. In the early 1990s, the crisis of the fishing industry resulted in both economic and human tragedies. Drawing conclusions from the crisis the fishing fleets experienced at that time, some groups of producers started to think of better enhancement of their production. Since then, they have tried to adapt their production to market demand. This has led them to set up quality initiatives (Brittany Quality Sea food, Brittany headland anglers, etc.), thus promoting their savoir-faire, the freshness certification, or the improvements in the quality of their products.

3. Translate the following passage into English.

海产品质量与安全问题日益突出,海产品食用风险的三个主要来源是化学危害因子、生物危害因子、海洋油污染。在此基础上结合国内外海产品食用安全政策,这篇文章对海产品食用风险提出了防范建议,包括加强海产品质量安全的法规建设,建立渔业环境和药物严格控制和管理机制;引入生命周期评价体系,加强海产品安全监测网络体系的建设,加强海域管理,加强海洋和海岛生态环境保护。

Unit 3
Analysis and Safety of Fishery Processing Market

Analysis of Fishery Processing Market

1. In recent years, there have been reports of seafood in the United States being labeled with an incorrect market name. FDA is aware that there may be economic **incentives** for some seafood producers and retailers to misrepresent the identity of the seafood species they sell to buyers and consumers. While seafood fraud is often an economic issue, species substitution can be a public health risk (e. g. , **substituting** a scombro-toxin- or ciguatoxin-associated fish for a non-toxin-associated fish).

2. For this reason, the Agency has invested in significant technical improvements to enhance its ability to identify seafood species using state-of-the-art DNA sequencing. DNA sequencing has greatly improved FDA's ability to identify misbranded finfish sea-food products in interstate commerce or offered for import into the United States. The Agency has trained and equipped eight field laboratories across the country to perform DNA testing as a matter of course for suspected cases of misbranding and for illness outbreaks due to finfish seafood, where the product's identity needs to be confirmed.

3. FDA also trained analysts from the U. S. Customs and Border Protection (CBP) and the National Marine Fisheries Service in its new DNA-based species identification **methodology**. FDA has made its **protocol** for using DNA sequencing for the identification of finfish products as well as its DNA reference standards publicly available through the FDA website. As a follow up to its now established capacity to identify finfish products using DNA, FDA has recently developed a protocol and a DNA reference library to extend these identification capabilities to include commercial species of shrimp, crab, and lobster. The Agency has already posted some of its DNA reference sequences for shrimp, crab, and lobster on its website and **anticipates** releasing the protocol to the public this year after final peer review, which will enable the seafood industry to monitor and test their products to confirm the species.

4. With DNA testing capacity in place, FDA has conducted DNA testing on fish that have a history of being misidentified in an effort to determine the accuracy of the market

names on their labels. These sampling efforts specifically targeted seafood reported to be at the highest risk for mislabeling and/or substitution, including cod, **haddock**, catfish, basa, swai, snapper, and grouper. As FDA announced in September 2014, the sampling and testing conducted as part of this project found that the fish species was correctly labeled 85 percent of the time. The Agency has the authority to take enforcement action against products in interstate commerce that are adulterated or misbranded and refuse admission of products imported or offered for import that appear to be adulterated or misbranded.

5. FDA will use the results from this testing to help guide future sampling, enforcement, and education efforts designed to ensure that seafood offered for sale in the U. S. market is labeled with an acceptable market name for the species. For instance, the Agency is conducting sampling and testing, in cooperation with state and local authorities, to look for mislabeling at the retail level. We also have posted on the FDA website a three-part learning module on proper seafood labeling to help the seafood industry, retailers, and state regulators ensure the proper labeling of seafood products offered for sale in the U. S. marketplace.

6. FDA's authority under the Federal Food, Drug, and Cosmetic Act (FD & C Act) provides a broad **statutory** framework to ensure that imported foods are safe, wholesome, and accurately labeled. It is the importer's responsibility to offer for entry into the United States product that is fully compliant with all applicable U. S. laws. Under the seafood HACCP regulation, HACCP controls are required for both domestic and foreign processors of fish and fishery products. Additionally, the regulation requires that U. S. importers take certain steps to verify that their foreign suppliers meet the requirements of the regulation.

7. As mentioned earlier, FDA uses a variety of measures to enforce processors' compliance with seafood HACCP, including inspections of foreign processing facilities, use of a screening system to sample imported products, domestic **surveillance** sampling of imported products, inspections of seafood importers, evaluations of filers of seafood products, foreign country program **assessments**, and relevant information from our foreign partners and FDA foreign office posts.

8. When an FDA-regulated product is offered for import into U. S. commerce, CBP procedures ensure that FDA is notified. If the product appears to be adulterated or misbranded, based on examination or other information, such as prior history of the product, **manufacturer**, or country, FDA will give notice advising the owner or consignee of the appearance of a violation under the FD & C Act and the right to provide **testimony** or evidence (such as a laboratory analysis by an independent laboratory) to rebut the appearance of the violation. In some circumstances, importers may request permission to recondition the product to bring it into compliance with applicable

requirements and regulations. If the product is ultimately refused admission, it must be destroyed, unless it is exported by the owner or consignee within 90 days of the date of the notice of refusal.

9. In 2002, the Congress gave FDA new authorities to enhance protection of the food supply in the Public Health Security and Bioterrorism Preparedness and Response Act. One of the most important provisions is the requirement that FDA be provided prior notice of food (including animal feed) that is imported or offered for import into the United States. This advance information enables FDA, working closely with CBP, to more effectively target food that may be intentionally **contaminated** with a biological or chemical agent or which may pose a significant health risk to the American public. Suspect shipments then can be intercepted before they arrive in the United States and held for further evaluation. To enhance targeting efforts on commercial imports, FDA participates in the Commercial Targeting and Analysis Center, which consists of CBP and nine other participating Federal agencies.

10. FDA has numerous other tools and authorities that enable the Agency to take appropriate action regarding imported products. In recent years, the Agency has significantly increased the number of inspections of foreign food manufacturers. For example, FDA conducted 1,336 foreign food facility inspections in FY 2014, compared to 153 inspections in 2008. Looking specifically at seafood, the Agency conducted 303 foreign seafood facility inspections in FY 2014, compared to 95 inspections in 2008. Furthermore, FSMA gave FDA the authority to refuse admission into the United States of food from a foreign facility, if FDA is refused entry by the facility or the country in which the facility is located upon FDA's request to inspect such facility.

11. Besides physical inspections of domestic and foreign facilities, the Agency's field force also conducts surveillance of food offered for import at the border to check for compliance with U. S. requirements. As part of our surveillance work at the border, FDA **utilizes** a risk-based approach to allocate resources, with priority given to high-risk food safety issues. FDA screens all import entries electronically prior to the products' entering the country, and a subset of those are physically inspected at varying rates, depending on the **potential** risk associated with them. Based on the risk ranking, the Agency will direct resources to the more critical activities that have a greater impact on public health. In FY 2014, FDA processed **approximately** 938,000 entries of imported seafood, while our field staff performed nearly 26,000 physical examinations of seafood imports and collected over 5,600 samples of domestic and imported seafood for analysis at FDA field laboratories.

(1,241 words)

➤ *New Words*

incentive [ɪnˈsentɪv]   *n.* something that encourages you to work harder, start a new activity, etc. 动机；刺激

If they have a direct incentive to do so they will think about it.

如果他们有直接的动机这样做,他们会考虑它。

substitute [ˈsʌbstɪtjuːt]   *v.* to use something new or different instead of something else 用(新的或不同的事物)代替

The recipe says you can substitute yoghurt for the sour cream.

食谱上说可以用酸奶代替酸味奶油。

methodology [meθəˈdɒlədʒɪ]   *n.* the system of methods followed in a particular discipline 方法学,方法论

We turn from methodology and science to politics.

我们从方法论和科学转变到政治。

protocol [ˈprəʊtəkɒl]   *n.* an international agreement between two or more countries; a written record of a formal or international agreement, or an early form of an agreement 协议；(协议的)文本,草案

Have them all use the same communications protocol.

让他们都使用相同的通信协议。

anticipate [ænˈtɪsɪpeɪt]   *v.* to regard something as probable or likely 预期,期望

Sales are better than anticipated.

销量比预期要好。

haddock [ˈhædək]   *n.* important food fish on both sides of the Atlantic; related to cod but usually smaller (鱼类)黑斑鳕,黑线鳕

This has benefited species such as American plaice, pollock, haddock and Atlantic cod.

美国鲽鱼、狭鳕、黑线鳕和大西洋鳕鱼因此受益。

statutory [ˈtʃətrɪ]   *a.* relating to or created by statutes 法定的；法令的；可依法惩处的

The education secretary had set out statutory guidelines for schools and councils to follow, he added, which would ensure that "inappropriate" content would not be used.

教育部长已经为学校和委员会设立了应该遵循的法定方针,他补充说,这可以确保不合适的内容不会被使用。

surveillance [sɜːˈveɪl(ə)ns]   *n.* close observation of a person or group (usually by the police) 监督;监视

So I think surveillance for these kinds of things should be increased.

所以我认为对这类事情的监视应该有所提升。

assessment [ə'sesmənt]   *n.* the classification of someone or something with respect to its worth 评定;估价

Rational, unemotional, self assessment should tell us whether or not we have the skills and ability to do something.

理性的、不掺杂感情的自我评价应该能告诉我们自己是否有能力和技巧去做些什么。

manufacturer [ˌmænjʊ'fæktʃərər]   *n.* a business engaged in manufacturing some product 制造商;厂商

They said themselves: "We are the best manufacturer of mirrors in America."

他们告诉自己:"我们是美国最好的主镜制造商。"

testimony ['testɪmənɪ]   *n.* a thing that shows that something else exists or is true 证据;证明

The increase in exports bears testimony to the successes of industry.

出口增长证明了产业的成功。

contaminate [kən'tæmɪneɪt]   *v.* to make impure 污染,弄脏

Phones and computers contain dangerous metals like lead, cadmium and mercury, which can contaminate the air and water when those products are dumped.

手机和电脑含有像铅、镉和汞这样的有毒金属,这些产品在被抛弃后,会污染水和空气。

utilize ['juːtˌlaɪz]   *v.* to put into service; to make work or employ (something) for a particular purpose or for its inherent or natural purpose 利用

If you plan to develop or execute using either of these scripting tools, you will need to utilize the SDP platform.

如果您计划使用这些脚本工具进行开发或者执行的话,那么您将需要利用这个 SDP 平台。

potential [pə'tenʃl]   *a.* that can develop into something or be developed in the future 潜在的;可能的

First we need to identify actual and potential problems.

首先,我们需要弄清实际的问题和潜在的问题。

approximately [ə'prɒksɪmətlɪ]   *ad.* (of quantities) imprecise but fairly close to correct 大约,

近似；近于

The yolk contains all the fat and approximately half of the protein of the egg.

蛋黄含有鸡蛋中全部的脂肪和大约一半的蛋白质。

➤ *Phrases and Expressions*

take action against 采取行动防止
be compliant with 符合；顺从
give notice 通知
in some circumstances 在某些情况下
compared to 与……相比

➤ *Translation*

1. FDA is aware that there may be economic incentives for some seafood producers and retailers to misrepresent the identity of the seafood species they sell to buyers and consumers. （Para. 1）
   美国食品及药物管理局意识到，一些海产品生产商和零售商可能会出于经济动机来歪曲他们向买家和消费者销售的海鲜品种的身份。

2. The Agency has trained and equipped eight field laboratories across the country to perform DNA testing as a matter of course for suspected cases of misbranding and for illness outbreaks due to finfish seafood, where the product's identity needs to be confirmed. （Para. 2）
   该机构已经在全国范围内培训和配备了八个实地实验室，对疑似病例和因食用有鳍鱼海产品引起的疾病进行 DNA 检测，因为这些海产品的身份需要得到确认。

3. FDA also trained analysts from the U. S. Customs and Border Protection （CBP） and the National Marine Fisheries Service in its new DNA-based species identification methodology. （Para. 3）
   FDA 还对来自美国海关与边境保护局（CBP）和国家海洋渔业局的分析人员进行了新的基于 DNA 的物种鉴定方法的培训。

4. With DNA testing capacity in place, FDA has conducted DNA testing on fish that have a history of being misidentified in an effort to determine the accuracy of the market names on their labels. （Para. 4）
   随着 DNA 检测能力的具备，FDA 已经对曾经被错误标识的鱼类进行了 DNA 检测，以确定其在市场上销售的标签名称的准确性。
   【独立主格结构表示时间、原因、条件。翻译的时候，可以根据具体的情况增加连词"因为……""由于……""……之后""当……的时候""如果……""假若……""只要……"等，然后译成相应的状语从句形式。】

5. We also have posted on the FDA website a three-part learning module on proper seafood labe-

ling to help the seafood industry, retailers, and state regulators ensure the proper labeling of seafood products offered for sale in the U. S. marketplace. (Para. 5)

我们还在 FDA 网站上发布了一个关于海产品标签的三部分学习模块,以帮助海鲜产业、零售商和国家监管机构确保在美国市场上出售的海鲜产品被正确地贴上标签。

6. It is the importer's responsibility to offer for entry into the United States product that is fully compliant with all applicable U. S. laws. (Para. 6)

进口商有责任提供完全符合美国所有适用法律的产品。

7. If the product is ultimately refused admission, it must be destroyed, unless it is exported by the owner or consignee within 90 days of the date of the notice of refusal. (Para. 8)

如果产品最终被拒绝进入,它必须被销毁,除非货主或收货人在收到拒绝入内的通知之日起的 90 天内把它运走。

8. In 2002, the Congress gave FDA new authorities to enhance protection of the food supply in the Public Health Security and Bioterrorism Preparedness and Response Act. (Para. 9)

2002 年,国会给予 FDA 新的权力,以增强公共卫生安全和生物恐怖主义防范与应对法案对食品供应的保护。

9. FDA has numerous other tools and authorities that enable the Agency to take appropriate action regarding imported products. (Para. 10)

FDA 有许多其他工具和职权,使该机构能够对进口产品采取适当的行动。

10. Besides physical inspections of domestic and foreign facilities, the Agency's field force also conducts surveillance of food offered for import at the border to check for compliance with U. S. requirements. (Para. 11)

除了对国内外设施进行实物检查之外,本机构外勤部门还对边境进口食品进行监测,以检查是否符合美国的要求。

【对于较长的句子,尤其是介词太多或从句太繁杂的句子,与其让人读得喘不上气儿,不如见招拆招,让句子更为利落。】

 **Exercises**

1. Fill in the blanks with the proper given words, and then translate the sentences into Chinese.

approximately	utilize	incentive	substitute
protocol	anticipate	assessment	take action against
potential	be compliant with	give notice	contaminate
in some circumstances	methodology	testimony	

1) If they have a direct _____ to do so they will think about it.

2) Emotion should never be a _____ for sound policy.

3) We turn from _____ and science to politics.

4) Have them all use the same communications _____.

5) So, even without understanding all that this file does, you can _____ the next step.

6) Every couple goes through a stage of _____ as you figure out if you want to be together: Do you want to live together?

7) His _____ contradicted that of the preceding witness.

8) Critics fear that fracking can _____ ground and surface water or even cause gas to leak into domestic water supplies.

9) We must _____ all available resources.

10) So, that tells me what the _____ should be here.

11) On some systems the time is a true count of milliseconds, changing _____ every millisecond.

12) What's at stake in that fight, above all, is the question of whether we'll _____ climate change before it's utterly too late.

13) Each product has its own different implementation technology and should _____ the WCAG Checklist.

14) The entry of the caution would freeze the register during its currency; and a caveat to _____ of a claim.

15) However _____ people consider the environmental effects as well.

2. Translate the following passage into Chinese.

With DNA testing capacity in place, FDA has conducted DNA testing on fish that have a history of being misidentified in an effort to determine the accuracy of the market names on their labels. These sampling efforts specifically targeted seafood reported to be at the highest risk for mislabeling and/or substitution, including cod, haddock, catfish, basa, swai, snapper, and grouper.

As FDA announced in September 2014, the sampling and testing conducted as part of this project found that the fish species was correctly labeled 85 percent of the time. The Agency has the authority to take enforcement action against products in interstate commerce that are adulterated or misbranded and refuse admission of products imported or offered for import that appear to be adulterated or misbranded.

3. Translate the following passage into English.

多年来,我们为保护我国水生生物资源和水域生态环境做出了不懈的努力,取得了一定成效。海洋伏季休渔、长江禁渔期、海洋捕捞渔船控制、重点水域污染防治等一系列有利于维护渔业生态安全的制度和措施实施,对缓解水生生物资源和生态环境的压力起到了一定作用。但是,目前过度捕捞导致水产资源减少,水域污染导致水域生态环境恶化,部分水域呈现生态荒漠化趋势的问题,尚未得到根本的遏制。维护渔业生态安全已经成为一项重要而紧迫的任务。

Chapter 4

Marine Mechanical Engineering

Unit 1
Design and Manufacturing of Watercraft

History of High Speed Ship Development

"I wish to have no connection with any ship that does not sail fast, for I intend to go into harm's way."

John Paul Jones, 1778

1. It is almost impossible to provide a clear historical basis for all the many ways that "high speed" has been introduced into the marine world. Inventors have **tumbled** over themselves in **conceptualizing**, building, testing and otherwise trying out their designs. History has been constantly repeating itself as new designs appear that are frequently nothing more than a reappearance of some earlier design but have gained status because the earlier designs were either lost in the **patent** offices or abandoned by the original inventor for say, financial burdens or personal crises and other reasons. Sometimes the "new" design came into being simply because the "time was right" and the original inventor is lost to **antiquity**. In some cases, the speed improvement is **overshadowed** by the introduction of a new and unexpected technology. The "invention" of the Clipper Ship that was so successful in the 1840s was soon **eclipsed** by the invention and application of the steam engine for marine use. In these days of seeking "alternative energy sources" perhaps the use of sails may come back!

2. There are many books already written that document the ways that "high speed" has been introduced in the marine field and it is not the purpose here to repeat such **documentation**. Christopher Dawson's book, *A Quest for Speed at Sea* published in 1972 provides an excellent treatment of advances in sail, engine **propulsion**, hull form and the start of the "dynamic lift" ships such as **hydrofoils** and air cushion craft. Dawson shows how the conflicting requirements for speed, load carrying, seaworthiness endurance, economy and reliability have greatly influenced each of the various designs

from historical times to the present day. Although Dawson's book was published over 40 years ago, not much has changed in these conflicting requirements since that time. Frequently, in modern day developments, time and money has been spent on re-living this history.

3. Another respected historian is H. F. King, a noted former editor of (now **defunct**) *Flight International and Air Cushion Vehicles*. King in his 1966 book *Aeromarine Origins* documents the results of his research into many of the designs used today. It is worth quoting from the flyleaf of his book: "One of the most remarkable facts to emerge is that the hydrofoil boat appears not only to have been a British invention, dating back over a hundred years, but that British ideas and techniques were used by some of the earliest experimenters abroad. The planing boat, or **hydroplane**, is seen as the invention of a British clergyman, who experimented with rocket-propelled craft of this type about a century ago. This same man proposed fearsome "rocket-rams", and one of these, intended to weigh 140 tons, is described and illustrated. Such weapons were to "sweep away existing navies" and to "render war at sea no longer possible." A vessel with an air **lubricated** hull, of the type claimed by recent American inventors as something novel, is shown to have been in regular public service on the New York—New Jersey run over a century ago. Well before the First World War, the Wright brothers were experimenting not only with hydrofoils, but with air lubricated floating bodies also. The author (Dawson) undertakes the fullest enquiry to date into man's earliest consciousness of the phenomenon of "ground effect", upon which the air cushion craft of today depends."

4. Much of this current book is concerned with the application of the "air cushion" to ship design because enough evidence now exists that suggests that for the large **tonnage** ship class of consideration for oceanic use that is required to carry large tonnage payloads for either commercial or military use, variations of **hovercraft**, air cushion craft, surface effect ships or similar variants, will be the most likely solution. Hydrofoil ships in their various forms (surface piercing or fully submerged) will be more applicable for short haul routes and **littoral** waters use and less so for transoceanic use (although this last point deserves re-examination). Work on hydrofoils will be discussed as needed in this overall context. For more detailed treatment of hydrofoils and their history, the reader is referred to the excellent treatments by Capt Robert Johnson, Chapter V, *Hydrofoils* in the Special Edition of Naval Engineers Journal, *Modern Ships and Craft* published February 1985; and Bill Ellsworth's book, *Twenty*

Foilborne Years, published by DTNSRDC in 1986.

5. Leslie Hayward, in his book, *The History of Air Cushion Vehicles*, published in 1963 provides an insightful documentation of the various designs and hardware examples of vehicles that have employed the principle of air cushion technology dating back to 1716 and to the present day (of then 1963) that includes many of the foundation designs of today's craft. His in-depth research of designs and patents reveals that "not much is new under the sun" and that much of the success of today's craft owes as much to "being in the right place at the right time" as it does to new thought.

6. In the book, *Air Cushion Craft Development* by this author, published by the US Government Printing Office in 1980, documents in some detail the major developments of the past using the above references **liberally** and takes the reader further into the technology of air cushion craft generally and in the technical detail of Surface Effect Ship (SES) developments around the world and specifically in the development of the high speed SES developments by the US Navy.

7. In the two decades (1960—1980) since the major thrusts in the US and around the world, the developments have slowed down considerably (in hydrofoil, hovercraft and SES development) and it is with this wealth of historical data base as "ground zero" that a perspective can now be gained as to issues, successes and failures that have occurred in this quest for high speed at sea. This book will pull heavily on the wealth of previous work to provide a useful perspective for further development.

(1,077 words)

➢ *New Words*

tumble ['tʌmbl]   *v.* to fall down, as if collapsing 摔倒

I shove forward one more time and, incredibly, the slab comes loose, and I tumble forward over it, caught in my own momentum.

我再一次向前猛推,难以置信的是,石碑松了,我向前跌倒在它上面,我自己的冲力使我摔倒。

conceptualize [kən'septʃuəlaɪz]   *v.* to have the idea for 使概念化

How we conceptualize things has a lot to do with what we feel.

我们如何使事物概念化与我们感觉到的东西密切相关。

patent ['pæt(ə)nt]   *n.* a document granting an inventor sole rights to an invention 专利权;专利证

And despite the flaws in the new laws, the attention to patent reform on both sides of the Atlantic is good news for business.

尽管新的法律中有缺陷,但是对企业来说分处大西洋两岸的欧美对专利改革的关注是个好消息。

antiquity [æn'tɪkwətɪ]   *n.* extreme oldness 高龄;古物;古代

Conquest and exile were events that normally would spell the end of a particular ethnic national group, particularly in antiquity.

一般情况下,一个种族群体的结局,不外乎就是征服其他民族或被其他民族征服,然后被流放,特别是在古代。

overshadow [ˌəʊvəˈʃædəʊ]   *v.* to exceed in importance; outweigh 使失色;使蒙上阴影

Still, these nagging problems should not overshadow the dramatic progress that women have made in recent decades.

尽管如此,这些冗繁的问题不应该掩盖女性在近十年里取得的巨大的进步。

eclipse [ɪ'klɪps]   *v.* to exceed in importance; outweigh 遮住……的光;使黯然失色

Mo did not want to be ruled by anyone and it is notable that she never allowed the men in her life to eclipse her.

莫不想被任何人管制，很显然，她从不允许她生活中的男性使她黯然失色。

documentation ［ˌdɒkjumenˈteɪʃn］   *n.* confirmation that some fact or statement is true 文件记载；文件证据；文献记录

When we have serious problems, they are usually in the area of documentation and help from hardware manufacturers.

如果说我们有严重的问题，那就是硬件开发商提供的文档和帮助不够多。

propulsion ［prəˈpʌlʃn］   *n.* the power that moves something in a forward direction 推进力

Due to the low cost of the battery cells and excellent performance at ambient temperature, Lithium-ion（Li-ion）battery is a promising technology for propulsion applications.

由于锂电池的低成本和在室温下的良好性能，其将会在推进力应用的技术上有广泛的前途。

hydrofoil ［ˈhaɪdrəfɔɪl］   *n.* a speedboat that is equipped with winglike structures that lift it so that it skims the water at high speeds 水翼艇

Bell is also known for many other landmark inventions including his work in optical telecommunications, hydrofoils and aeronautics.

贝尔同样因为许多里程碑一样的发明而闻名，包括他在光纤通信、水翼艇技术和航空学方面的研究。

defunct ［dɪˈfʌŋkt］   *a.* having ceased to exist or live 非现存的；死的

Practical traders, who believe themselves to be quite exempt from any intellectual influences, are usually slaves of some defunct mathematicians.

讲求实际的交易者自信能免受任何智者的影响，却常常沦为一些已经作古的数学家的奴隶。

hydroplane ［ˈhaɪdrəpleɪn］   *n.* a speedboat that is equipped with winglike structures that lift it so that it skims the water at high speeds 水上飞机

It was the first time he had called on me, though I had gone to two of his parties, mounted in his hydroplane, and, at his urgent invitation, made frequent use of his beach.

这是他第一次来看我,虽然我已经赴过两次他的晚会,乘过他的水上飞机,而且在他热情邀请之下时常借用他的海滩。

lubricated ['lu:brɪkeɪtɪd]　　*a.* smeared with oil or grease to reduce friction 润滑的

One analysis suggests that as water melts from the ice cap, the glaciers are lubricated and rush along the rock more quickly, on their way to tipping into the sea.

一项分析显示,水从冰帽上融化时,就润滑了这些冰川,这些冰川沿着石头下滑得更快,最终翻倒在海中。

tonnage ['tʌnɪdʒ]　　*n.* the total number of tons that something weighs 吨位;吨数

They also want to almost double the tonnage of ships the port currently handles.

他们还希望把这个港口目前的船只吞吐量吨位提高一倍。

hovercraft ['hɒvəkrɑːft]　　*n.* a craft capable of moving over water or land on a cushion of air created by jet engines 气垫船

Travelling at speeds of up to thirty-five knots, these hovercraft can easily outpace most boats.

这些气垫船以高达每小时 35 节的速度航行,很容易超过大部分船只。

littoral ['lɪtərəl]　　*a.* of or relating to a coastal or shore region 沿海的;海滨的

The aircraft is designed to provide battle-group protection and add significant capability during coastal, littoral and regional conflicts.

这种飞机为战场提供地面防护,并显著增强了在沿海、濒海和区域冲突中的能力。

liberally ['lɪbərəlɪ]　　*ad.* in a generous manner 充足地,大量地;自由地

Some writers, however, from sheer exuberance or a desire to show off, sprinkle their work liberally with foreign expressions, with no regard for the reader's comfort.

但是,有的写作者或纯粹是为了丰富词汇表达,或是为了炫耀,在自己的作品中遍撒外来语,全不顾读者的感受。

➤ *Phrases and Expressions*

come into being 形成；产生

date back 追溯到；回溯至

experiment with 进行实验；试用

to date 至今；迄今为止

undertake the enquiry into 对······进行调查

owe much to 多亏了；在很大程度上归功于

➤ *Terminology*

Clipper Ship 高速帆船；古时快帆船

dynamic lift 动力举升；动升力

rocket-propelled 用火箭推进的

air cushion 气垫

payload（导弹、火箭等的）有效载荷

short haul route 短程航线

Surface Effect Ship 表面效应船

➤ *Proper Names*

New Jersey 美国新泽西州

the US Government Printing Office 美国政府印刷局

➤ *Translation*

1. It is almost impossible to provide... introduced into the marine world. （Para. 1）
   对于将"高速"引入海洋世界的所有方式，要提供清晰的历史依据，几乎是不可能的。

2. Sometimes the "new" design came... original inventor is lost to antiquity. （Para. 1）
   有时，"新"设计的出现仅仅是因为"时机成熟"，而最初的发明者已经不合时宜。
   【在进行翻译时，词典的帮助起着很大的作用。在几本词典中查 antiquity 这个词，可以得到"高龄；古代；古物；古代的遗物"等几个释义，但这几个释义放在本句中都不合适，所以光靠词典的帮助，而不借助具体的上下文是不够的。本句上文说"新"设计的出现仅仅是因为"时机成熟"，所以下文意译为"最初的发明者已经不合时宜"。】

3. There are many books... purpose here to repeat such documentation. （Para. 2）
   已经有很多书记录了高速技术在海洋领域的应用，这里就不再重复这些记录了。

4. Dawson shows how the conflicting requirements... to the present day. （Para. 2）
   人们对速度、载重、适航耐力、经济性和可靠性的要求是相互矛盾的。道森展示了从历史时

期到现在,这些要求如何极大地影响了各种设计。

5. One of the most remarkable facts... the earliest experimenters abroad.（Para. 3）
最值得注意的事实之一是,水翼艇似乎不仅是一百多年前英国人的发明,而且英国的思想和技术也被国外最早的一些实验人员使用过。

6. A vessel with an air lubricated hull... New Jersey run over a century ago.（Para. 3）
最近美国发明家们把这种具有空气润滑船体的船称为新奇之物,其实一个多世纪以前,这种船曾在纽约—新泽西线上进行过常规的公共服务。
【在英语词汇里,run 一词语义最为丰富,在上海译文出版社 1991 年出版的《英汉大词典》里,run 作为名词有 38 个释义,这就需要我们依靠具体的上下文来判断它的意义。在"on the New York—New Jersey run"这个短语中,run 作为名词接近"奔跑;赛跑;趋向;奔跑的路程"这几个释义,根据上下文选取"奔跑的路程"这个释义,引申为船只的航线,所以这个短语译为"在纽约—新泽西线上"。】

7. The author（Dawson）undertakes... air cushion craft of today depends.（Para. 3）
关于人类对"地面效应"现象最初的了解,作者(道森)进行了迄今为止最全面的调查,而当今的气垫船正是依靠这种效应。

8. Leslie Hayward, in his book... foundation designs of today's craft.（Para. 5）
莱斯利·海沃德在他 1963 年出版的《气垫交通工具的历史》一书中,对交通工具的各种设计和硬件实例进行了详尽的记录,这些交通工具采用的气垫技术原理可以追溯到 1716 年至当时(当时为 1963 年),其中包括当今船只的许多基础设计。

9. His in-depth research of designs and patents... as it does to new thought.（Para. 5）
他对设计和专利的深入研究表明,"世界上没有什么是新的",当今船只的成功在很大程度上要归功于"出现在恰当的时间和恰当的地点",也要归功于新的思想。

10. In the two decades（1960—1980）... quest for high speed at sea.（Para. 7）
在美国和世界重大攻势过后的二十年(1960—1980)里,(水翼艇、气垫船和表面效应船方面的)发展速度大大减慢,而凭借如此丰富的历史数据库作为"起点",现在人们可以对寻求海上高速航行中出现的问题、成功和失败形成看法。

 **Exercises**

1. Fill in the blanks with the proper given words, and then translate the sentences into Chinese.

provide	constantly	original	design
treatment	lift	various	abroad
owe	cushion	reveal	regular
spend	frequently	under	craft

1) A survey of the American diet has _____ that a growing number of people are overweight.

2) He was so engrossed in his work that he _____ stayed up until the small hours.

3) In doing translation, one should not alter the meaning of the _____ to suit one's own taste.

4) We aren't very wealthy, so we don't have enough money to _____ on that even if I wanted to.

5) There is also a new steering wheel with an energy-absorbing rim to _____ the driver's head in the worst impacts.

6) And now we aim to try to help _____ new treatment for these severely disabled patients.

7) _____ the new regulations, one in five cars may need repairs costing as much as $120.

8) They'll have something going and then almost unbeknownst to you they will take that and _____ it up in terms of the pitch content.

9) When Bernardello asked them to build him a home, they drew up the _____ in a week.

10) I _____ a debt of thanks to Joyce Thompson, whose careful and able research was of great help.

11) And our engineers _____ hear from other friends working for other Internet companies in Beijing—they live in the flashiest office space in Beijing and their salary is raised three times a year.

12) Gone are the days in which we can apprentice our way through the years until we become the master of our _____, sharing our carefully honed techniques with the next generation.

13) How many chords are involved in this chord change and are they changing at a _____ or irregular rate?

14) Many patients are not getting the medical _____ they need.

15) I have watched them go through all of the _____ stages over the years and their planning and budgeting has been stellar.

16) Another aspect, though, is that the US invested its money _____ at high returns, while the rest of the world invested in the US at low returns.

2. Translate the following passage into Chinese.

Much of this current book is concerned with the application of the "air cushion" to ship design because enough evidence now exists that suggests that for the large tonnage ship class of consideration for oceanic use that is required to carry large tonnage payloads for either commercial or military use, variations of hovercraft, air cushion craft, surface effect ships or similar variants, will be the most likely solution. Hydrofoil ships in their various forms (surface piercing or fully submerged) will be more applicable for short haul routes and littoral waters use and less so for transoceanic use (although this last point deserves re-examination). Work on hydrofoils will be discussed as needed in this overall context.

3. Translate the following passage into English.

高速船舶的设计原理远不如那些经过数百年实践磨炼的、起源于排水船体的低速船舶的设计技术完善。另一方面,高速船舶设计涉及不同船型概念的研究,每一个概念都需要了解导致船舶升力和阻力的四种基本力。这四种力是:流体静力(浮力)、流体动力、空气静力和空气动力。这四种力按不同的物理定律进行缩放,使得从小型模型到大型船舶的放大变得困难。这些力的组合在每种情况下均会有所不同,具体取决于操作者所考虑的概念选择。

Unit 2
Marine Energy and Power

Waves as Energy Carriers

1. Who can contemplate the surface of the ocean and not notice its **incessant**, seemingly random motion? This motion represents the net effect of long swells bearing memories of distant storms, shorter waves driven up by winds blowing nearby and over the surrounding area, and very small ripples caused by surface tension, with the pattern constantly changing and its **constituents** continuously traveling in different directions. Closer to shore, the ocean surface seems dominated by waves of steeper slopes with aligned crests mostly traveling towards the shore. As the crests approach the shore they get steeper, slightly disintegrating near their tops but still getting taller until they turn into plunging, roaring breakers, finally giving up their energy as they **dissipate** (with some reluctance) over sandy beaches or rocky shores. Whether **maneuvering** (or simply observing) a small craft on its way through the waves, or watching the waves crash loudly onto the beach, one can hardly fail to be impressed by their **unrelenting** display of energy.

2. Waves carry energy from one point to another. One observes an excellent illustration of this phenomenon when the relatively calm surface of a pond or a small lake is disturbed by a pebble or a small stone thrown into the water. A pattern of waves **emanates** from where the stone fell, traveling **radially** out and creating ever-expanding circles with alternating crests and troughs that get smaller, the farther the disturbance travels. In this case, waves carry away a portion of the kinetic energy of the stone over the water surface. A roughly similar phenomenon occurs when wind blows over a region of the ocean surface and generates waves that are higher and longer, the faster and longer the wind blows. In deep waters, the dissipation rate of the energy contained in waves is very small, enabling them to last longer and travel farther than the wind that created them. This happens for wind blowing over different regions, leading

to an averaging, storing, and concentration of energy, both spatially and temporally. Because the density of sea water is about three orders of magnitude greater than that of air, energy delivered by waves is also more concentrated than wind energy. Some waves grown by distant and local wind action get too large to sustain their forms as they drive themselves to the coast over the diverse shapes and features of the sea floor. Large waves will lose energy to **friction** with the sea floor and ultimately break on the way to and on the shore. Their continual pounding action on the shore can break rocks and lead to beach formation and sometimes erosion of the coast line.

3. An interesting aspect of wave motion is that, although the energy **propagates** along the sea surface (the "free surface"), the surface itself only **undulates**, with an up-down motion that is a function of location and time. Distilled down to the most basic description, the most straightforward wave to think about is a **sinusoidal** train of peaks ("crests") and valleys ("troughs") that travel in a single direction. For such a wave, while the water surface **oscillates** up and down, the water particles under the surface perform a circular motion about a central mean position that remains unaltered in time. The top of the particle's circular **trajectory** coincides with the wave crest, while the bottom defines the wave trough. The frequency of a wave equals the angular velocity of particle motion, and the **amplitude** equals the **radius** of particle motion. The phases of various particle motions in the direction of wave propagation are such that the distance between successive peaks (troughs), or wavelength, is related to the frequency through the so-called dispersion relation, which in deep water also contains gravity acceleration. After all, it is really the restoring action of gravity to counter any disturbing action of wind (or any other surface forcing) that sustains the waves and water particle motion. In shallower water, water depth also becomes increasingly important and plays a prominent role in the dispersion relation. Particle motion becomes increasingly **elliptical**, with the horizontal **excursion** progressively exceeding the vertical.

4. A compact mathematical description of the relation between particle motion and water surface **deflection** is possible if one ignores the relatively small contribution of **viscous** effects affecting wave motion. Water is for most purposes **incompressible** and the particles travel without deforming in shear, so methods of ideal fluid hydrodynamics are by and large sufficient. Specifically, the particle velocity vector components under the wave can be defined as spatial derivatives (along the three coordinate directions) of a **scalar** "velocity potential". The water surface "elevation" or "wave profile" (more accurately referred to as "free surface elevation") changes at a rate such that no individual particles leap out of or dive into the surface. More precisely, the rate of change of surface elevation equals the vertical velocity of individual particles over the free surface. The time rate of change of surface elevation thus equals the **derivative** of the velocity potential (at any given point) **perpendicular** to the free surface. Clearly, such a representation cannot be used to model sprays produced when a wave strikes a structure or the actual dynamics near the top of a breaking wave.

5. Wave forces and moments on the device can then be found using the velocity potential, which can be determined analytically or, more commonly, numerically, such that it satisfies the boundary conditions at the device. When the device is able to oscillate in response to the waves, its oscillation velocity determines the boundary conditions on it. When it is held stationary, the fact that fluid cannot flow across the device walls provides the necessary boundary conditions. Solving for the velocity potential in the presence of the device enables one to solve for the device oscillations and forces/moments, and then to use these solutions to determine device's energy conversion abilities.

(1,000 words)

➤ *New Words*

incessant [ɪn'sesnt]	*a.* uninterrupted in time and indefinitely long continuing 不断的;连续的
	He did grow frustrated with the tedious nature of the incessant questions and constant discussions on the matter.
	关于这个问题,没完没了的提问和没完没了的讨论实在乏味,令他沮丧。
constituent [kən'stɪtʃuənt]	*n.* an artifact that is one of the individual parts of which a composite entity is made up 成分
	Nutrients get broken down into several classes and these are the constituents of food that your body uses for one purpose or another.
	营养素被分为几个类别,它们是食物的组成要素,支持身体机能运转。
dissipate ['dɪsɪpeɪt]	*v.* to move away from each other 消散;散逸
	A full meltdown at the Japanese facility would still release radioactive gases, but those tend to dissipate in the atmosphere.
	日本这一核电站的核芯全部熔解仍将会释放出放射性气体,这些气体往往会消散在大气中。
maneuver [mə'nʊvə]	*v.* to direct the course; determine the direction of travelling 调遣;操纵
	You'll have to help, since he won't have the dexterity to successfully maneuver a toothbrush.
	你得帮助他,毕竟他的手指不够灵巧还不能很好地使用牙刷。
unrelenting [ˌʌnrɪ'lentɪŋ]	*a.* continuing without stopping 无休止的
	Survivors also aided in the search effort, but after two days with little food or water and unrelenting heat, more and more residents are beginning to give up hope.
	幸存者也参与了搜索工作,两天过去了,没有多少食物和水再加上持续高温,越来越多的居民开始放弃希望。
emanate ['emǝneɪt]	*v.* to proceed or issue forth, as from a source 产生;散发
	The waves that emanate from the collision of two black holes should be detectable.
	两个黑洞相撞产生的波应该是可测。
radially ['reɪdɪǝlɪ]	*ad.* in a radial manner 放射状地

So they migrate radially—in all possible directions away from the den—like starburst embers from a Fourth of July rocket.

于是,它们呈放射状移动——从巢穴向所有可能的方向迁移——像国庆焰火似的四射开来。

friction [ˈfrɪkʃn]　　　　　*n.* the force that makes it difficult for things to move freely when they are touching each other 摩擦;摩擦力

In most machines friction consumes effort, with the result that less work is got out than is put in.

在大多数机器中,摩擦要消耗功,以致造成输出的功要小于输入的功。

propagate [ˈprɒpəgeɪt]　　*v.* to become distributed or widespread 传播

The sending server does not propagate the password or any other private credentials (secret).

发送服务器并不传播密码或者任何其他私有的凭据(秘密)。

undulate [ˈʌndjuleɪt]　　　*v.* to move in a wavy pattern or with a rising and falling motion 起伏;波动

As we travel south, the countryside begins to undulate as the rolling hills sweep down to the riverbanks.

随着我们向南行进,乡间的山丘变得连绵起伏与河岸相接。

sinusoidal [ˌsaɪnəˈsɔɪdəl]　　*a.* of or relating to a sine curve 正弦曲线的

Water is attached to the body by surface tension, they thought, and the sinusoidal shaking creates centripetal forces that ejects that water off the body.

他们认为水沾在身上靠的是身体表面的亲和力,正弦波状的抖动动作制造了水脱离身体的离心力。

oscillate [ˈɒsɪleɪt]　　　　*v.* to move or swing from side to side regularly 振荡;摆动

You'll notice that at the line where the blue and red meet your eye seems to oscillate back and forth between the two.

你会注意到,当蓝色和红色相遇的时候,你的眼球就会在这两个颜色之间来回摆动。

trajectory [trəˈdʒektərɪ]　　*n.* the path followed by an object moving through space 轨道;轨线

In his theory of general relativity, Einstein realized that space and time can stretch and warp in ways that change the trajectory of light.

爱因斯坦在他的广义相对论中认识到,可以以改变光轨迹的方式对时间和空间进行拉伸和弯曲。

amplitude ['æmplɪtjuːd]
n. the maximum displacement of a periodic wave 振幅
When we measure its momentum, we are treating it as a wave, meaning we can know the amplitude of its wavelength but not its location.
当我们测量其动量时,我们将其当作波,这意味着我们能知道其波长的振幅,但无法获知其位置。

radius ['reɪdiəs]
n. the length of a line segment between the center and circumference of a circle or sphere 半径
To translate this formula into a expression, you would use variables to record the radius and circumference.
要将这个公式转换为表达式,你需要使用变量来记录半径及周长。

elliptical [ɪ'lɪptɪkl]
a. rounded like an egg 椭圆的
Unlike the elliptical galaxies, the spiral is rich in dust and gas for the formation of new stars.
与椭圆形星系不同,漩涡型星系含有丰富的尘埃和气体,这些都是形成新的恒星的原料。

excursion [ɪk'skɜːʃn]
n. wandering from the main path of a journey 偏移
Experimental results show that the proposed algorithm can track automatically and correct the excursion during tracking.
实验结果表明,本方法可以实现对目标的自动跟踪,同时有效修正了跟踪偏移。

deflection [dɪ'flekʃn]
n. the amount by which a propagating wave is bent 挠曲;挠度
Compared with traditional electrostatic cantilever actuators, this device can control the deflection of the beam more accurately.
与传统的静电悬臂梁执行器相比,该装置能更精确地控制梁的挠度。

viscous ['vɪskəs]
a. having a relatively high resistance to flow 黏性的;黏的
The more viscous lava is, the harder it is for gases within it to bubble out, so such lava has an explosive tendency.
岩浆越黏,其内部的气体想通过气泡冒出就越困难,所以这种岩浆有爆炸的倾向。

incompressible [ɪnkəm'presɪb(ə)l] *a.* impossible to compress 不能压缩的

Above the critical temperature and under high pressure, the vapor may become as dense and as incompressible as the liquid at lower temperature.

在临界温度以上和高压下,蒸汽密度可能变得像在低温下液体那样致密且不可压缩。

scalar [ˈskeɪlə(r)] 　*a.* of or relating to a directionless magnitude 标量的

Some metrics are scalar objects describing a single data point, such as the current firmware version on the appliance.

有些指标是描述单个数据点的标量对象,如设备上的当前固件版本。

derivative [dɪˈrɪvətɪv] 　*n.* the instantaneous change of one quantity relative to another 导数

Let's sneak in one more derivative here, which is to take the derivative of the derivative.

我们再求一次导数,也就是对导数求导。

perpendicular [ˌpɜːpənˈdɪkjələ(r)] 　*a.* intersecting at or forming right angles 垂直的

The tension is not doing any work, because the tension, which is in this direction, is always perpendicular to the direction of motion.

这个压力不做功,因为这个方向的压力总是和运动方向垂直。

➤ *Phrases and Expressions*

be impressed by 被……所感动;被……给予深刻印象
a portion of 一部分的
lead to 导致;通向
coincide with 巧合;同时发生;与……一致
after all 毕竟;终究
in response to 应答;对……有反应
by and large 大体上;总的来说

➤ *Terminology*

order of magnitude 数量级
crest 波峰
trough 波谷
dispersion relation 色散关系;频散关系
angular velocity 角速度

particle velocity vector 粒子速度矢量

spatial derivatives 空间导数

velocity potential 速度势

wave profile 波形;波剖面

free surface elevation 自由水面高度

➤ *Translation*

1. This motion represents the net effect... traveling in different directions. (Para. 1)

   这种运动表现了长涌浪、较短波浪和细小波纹的净效应。长涌浪承载着遥远风暴的记忆,较短波浪靠附近和周围地区吹来的风推动,细小波纹由水面张力引起。它们的模式不断变化,组成部分不断向不同的方向移动。

   【此句较长,英语长句的翻译首先要判断出句子的结构,再找出句中的主要句子成分,然后再分清句中的宾语、状语、表语、宾语补足语、定语等。此句属于简单句,只有一个主语 "This motion"、一个谓语"represents"、一个宾语"the net effect"和一个状语"with..."。其中宾语的修饰成分分为三层,分别是 of long swells、shorter waves 和 very small ripples。英语长句一般采取分译法。】

2. As the crests approach the shore... over sandy beaches or rocky shores. (Para. 1)

   当波峰接近海岸时,它们会变得更陡,在靠近顶部的地方会有轻微的崩解,但仍然会变得更高,直到它们变成向前猛冲的、咆哮的拍岸浪,并最终在沙滩或多岩石的海岸上消散(有些不情愿)时,耗尽了它们的能量。

3. A pattern of waves emanates... the farther the disturbance travels. (Para. 2)

   一种模式的波浪从石头落下的地方发散出来,呈放射状传播出去,并形成不断扩大的圆圈,这些圆圈的波峰和波谷相互交替,波峰和波谷越小,扰动传播的距离就越远。

4. Because the density of sea water... concentrated than wind energy. (Para. 2)

   由于海水的密度比空气的密度大 3 个数量级,所以海浪所传递的能量也比风能更集中。

5. Some waves grown by distant... shapes and features of the sea floor. (Para. 2)

   一些在远方来风和局地风作用下形成的波浪,越过海底各种不同的地形和地貌向岸边涌动,同时因为风势过大无法维持本来的形状。

   【英语具有直接性,往往是先表明结论再进行论证或描述事实,把句子的主要信息置于次要信息之前,即重心前置。汉语则习惯于按照事情的发展顺序,由事实到结论或由因到果进行论述,即重心后置。翻译时要根据译文的语言习惯,对原文的语序进行调整,使译文做到最大限度上的通顺,这就是换序译法。】

6. An interesting aspect of wave motion... function of location and time. (Para. 3)

   波浪运动的一个有趣之处是,虽然能量沿着海面(自由表面)传播,但海面本身只会随着位置和时间的变化而上下波动。

7. The phases of various particle motions... gravity acceleration. (Para. 3)

在波传播方向上,各种粒子运动的相位使得连续的峰(谷)之间的距离或波长通过所谓的色散关系与频率相关,该色散关系在深水中也包含重力加速度。

8. After all, it is really the restoring... waves and water particle motion. (Para. 3)

毕竟,真正维持波浪和水粒子运动的是重力的恢复作用,它抵御风的干扰作用(或任何其他表面张力作用)。

9. A compact mathematical description... effects affecting wave motion. (Para. 4)

如果忽略黏性效应对波浪运动相对较小的影响,则可以对粒子运动与水表面挠度之间的关系进行紧凑的数学描述。

10. Water is for most purposes incompressible... by and large sufficient. (Para. 4)

水在大多数情况下是不可压缩的,粒子在切变过程中运动而不变形,因此理想流体动力学方法大体上是足够适用的。

 **Exercises**

1. Fill in the blanks with the proper given words, and then translate the sentences into Chinese.

random	notice	net	surrounding
tension	steep	sustain	portion
region	feature	circular	deliver
depth	particle	density	prominent

1) Now you have this situation where you have a droplet that is _____ negative or perhaps net positive.

2) Every substance, no matter what it is, is composed of very small _____ called molecules.

3) In summer, its _____ mountainsides shimmer with wildflowers, and glacial rivers irrigate small valley farm fields and orchards, which yield generous crops of peas, potatoes, apples and plums.

4) He stressed that people should not hesitate to contact the police if they had _____ any strangers recently.

5) I was in the _____ of despair when the baby was terribly sick every day, and was losing weight.

6) Like all of you, I pray that the peoples of the _____ choose the path less travelled, the path of liberty.

7) Although we'd planned to have our baby at home, we never expected to _____ her ourselves!

8）When I have something I have to remember（a bill to pay in a few weeks for instance）I put it in a _____ place on the pin board above my computer.

9）We sat down, I upon a large, low divan, he with his back to the window and to a large _____ clock.

10）It's a bonus if you have moved to a new city with _____ roommates because you can easily join in on the things that they do.

11）The fear is that melting ice, along with increased snow and rain, could reduce the _____ and salinity of the top layers of the sea, making them more buoyant.

12）The cash dividends they get from the cash crop would _____ them during the lean season.

13）Whatever shapes they have, the mascots fundamentally show the distinctive geographical _____, history and culture unique to the host city.

14）As the cable wraps itself around the wheel, there is provision for adjusting the _____ of the cable.

15）The next day after the landing, they cut the _____ trees down and built a small house.

16）The homework _____ of your total grade is the aggregate of all of those weekly test scores.

2. Translate the following passage into Chinese.

Wave forces and moments on the device can then be found using the velocity potential, which can be determined analytically or, more commonly, numerically, such that it satisfies the boundary conditions at the device. When the device is able to oscillate in response to the waves, its oscillation velocity determines the boundary conditions on it. When it is held stationary, the fact that fluid cannot flow across the device walls provides the necessary boundary conditions. Solving for the velocity potential in the presence of the device enables one to solve for the device oscillations and forces/moments, and then to use these solutions to determine device's energy conversion abilities.

3. Translate the following passage into English.

线性系统模型确实适用于许多实际设备,而且常与许多弱非线性设备近似。线性系统模型的优点在于能够使用大部分可用于分析谐波的分析仪器。具体地说,利用线性系统模型,可以通过分析谐波的速度势,来理解设备对波的响应。这种分析设备对不规则波的响应的方法称为"频域分析"。频域分析本质上需要一定长度的波或振荡记录,因此不适用于实时逐波了解设备对不规则波的响应。对于动力可以视为线性的设备,为了描述作用于入射波表面信号或设备振荡的波力项,逐波工作需要脉冲响应核。

Unit 3
Marine Energy and Environmental System

Ocean Energy Technologies

1. The generation of clean electricity from the ocean's waves and tides has long been a dream. The existence and extent of the resource are without question, yet ocean energy has proven incredibly hard to realize. For decades, a wild and unforgiving sea seemed to place ocean energy just beyond our grasp. However, these technologies are now **tantalizingly** close to commercial deployment. The need for clean energy is urgent, and our maritime technologies have advanced in leaps and bounds. **Paradoxically**, the experience gained from exploiting offshore fossil fuels has helped us overcome many of the technical challenges of working in a hostile marine environment, and ocean energy is approaching a tipping point as arrays of devices begin to enter the water.

2. The development of renewable energy technologies and their integration into the energy mix will reduce **emissions**, conserve finite resources and increase energy security, in addition to bringing additional co-benefits, such as job creation. As a result, investment in renewable energy has grown dramatically in the last decade. The **imperative** to **decarbonise** the energy system initially drove the development of solar and onshore wind power, but interest has now spread offshore. The marine environment contains a number of potential sources of renewable energy, including wave and tidal energy (here collectively termed "ocean energy"), ocean thermal energy, salinity gradient or osmotic energy and offshore wind energy. Offshore wind in particular has seen rapid growth in recent years, while wave and tidal technologies are now also beginning to attract considerable interest and investment.

3. The oscillatory or circular motions of waves produced near the sea's surface can be converted into electricity, while tidal or marine **hydrokinetic** energy comprises ocean currents or tides that can be directly extracted and converted into electricity. There are a seemingly infinite number of ways that ocean energy technologies can harness the wave and tidal power of the oceans. The following quote from Stephen Salter, in-

ventor of the Salter's (Edinburgh) duck device for extracting wave energy, is indicative:

> Waves are only one of many possible sources and there are many possible ways in which waves can be harnessed. There are floats, flaps, ramps, funnels, cylinders, air-bags and liquid pistons. Devices can be at the surface, the sea bed or anywhere between. They can face backwards, forwards, sideways or **obliquely** and move in heave, surge, sway, pitch and roll. They can use oil, air, water, steam, gearing or electro-magnets for generation. They make a range of different demands on attachments to the sea bed and connections of power cables.

4. Ocean energy technologies were first explored in a number of countries as a result of the 1970s oil crisis. During this time, the UK's Department of Energy ran a wave energy programme aimed at upscaling prospective devices, while the United States focused on OTEC (ocean thermal energy conversion). However, this initial enthusiasm diminished in the face of high costs and the easing of the oil crisis, and subsequent support for ocean energy was **sporadic**. By the late 1980s, research activity had greatly diminished, and R&D funding was insignificant through much of the 1990s.

5. Climate change and energy concerns have brought a "second wave" of interest to ocean energy, particularly in the UK, Ireland, Portugal, Denmark, France, Australia, South Korea, Canada, and the United States. Large-scale utilities, energy agencies and industrial companies are making significant investments in the sector, and the European Marine Energy Centre (EMEC), a test bed for emerging ocean energy technologies, is fully subscribed with 14 full-scale devices generating to the grid. Similar testing centres are under development elsewhere. The military has also shown interest in supporting the development of ocean energy technologies, with a US naval base in Hawaii developing a wave energy generation project and a naval base in Western Australia signing an agreement to meet its power needs through ocean energy.

6. In August 2016, the first array of tidal turbines sent energy to the grid in the Shetland Islands, while in September the Orkney Islands saw the launch of the first of four 1.5 MW tidal turbines, the first phase of a project eventually intended to reach an installed capacity of 398 MW. Devices are advancing rapidly, costs are coming down, and commercialization appears to be on the horizon: "The technologies have been **optimized**, extensively tested and pilot wave energy projects have been realized. The knowledge and experiences gained lead to a development status that is ready for the market".

7. While recent developments in ocean energy are encouraging, a word of caution is advisable. Despite successful prototype deployments and the arrival of the first arrays and large-scale projects, ocean energy remains the least developed of the renewable energy technologies. Overall, ocean energy is perhaps at the same level of development as wind technology in the 1970s and early 1980s, when a range of wind turbine concepts were being researched and developed but uncertainty remained as to which concepts would become cost-competitive. Like many new industries, ocean energy **proponents** have often been overly ambitious or optimistic in their assessments of the future of ocean energy, and they risk overselling the potential or pace of the industry. At the same time, a wave of recent bankruptcies and technology failures, though perhaps inevitable, have shaken the industry and cast doubt on its ability to deliver.

8. A similar history plagued OTEC. There was a surge of interest in in the 1970s, but the predictions did not come to fruition as interest **waned**. In 1980, it was thought that research in key countries had advanced the technology to a stage where commercial OTEC devices could be deployed before the end of the decade and that, by the turn of the century, OTEC would be a commercial source of energy. Hearings before the House Subcommittee on Oceanography in the United States in 1978 estimated that OTEC generation would produce 3% of US energy requirements by 2000. In spite of considerable R& D efforts into the technology and the efforts of academics and legislators to improve the **regulatory** regime, OTEC did not ultimately live up to these expectations. While OTEC's **demise** was largely due to the easing of the oil crisis, in contrast to the more **intractable** climate and energy issues now faced by policymakers, this nonetheless cautions against making **unduly** optimistic assessments of the potential of ocean energy technologies.

(1,065 words)

➤ *New Words*

tantalizingly [ˈtæntəˌlaɪzɪŋlɪ]　　*ad.* in a tantalizing manner 令人着急地

While the answer is not yet known for certain, it appears to be tantalizingly close to the critical density.

虽然并没有确切的答案,但它似乎已经非常接近临界密度。

paradoxically [ˌpærəˈdɒksɪklɪ]　　*ad.* in a paradoxical manner 自相矛盾地;似是而非地;反常地

Paradoxically, Americans today seem less interested in the wider world than they were before the Twin Towers were felled.

反常的是,美国人如今对于世界其他地区的兴趣相比双子塔倒掉之前大大减弱了。

emission [ɪˈmɪʃn]　　*n.* the release of something into the atmosphere 排放

The emission of gases such as carbon dioxide should be stabilized at their present level.

二氧化碳等气体的排放应稳定在目前的水平。

imperative [ɪmˈperətɪv]　　*n.* something that is extremely important and must be done 紧迫之事

College is imperative for a long and successful life and can make your dreams come true in the long run.

对于长久和成功的人生来说,大学是必要的,从长远看,它能让你梦想成真。

decarbonise [ˌdiːˈkɑːbənaɪz]　　*v.* to remove carbon from (an engine) 脱碳;去碳

Special solvents can be used to decarbonise and desulfurize waste catalysts to replace the high-temperature roasting pretreatment process.

可以采用特殊的溶剂对废催化剂进行脱碳和脱硫预处理,以代替高温焙烧预处理。

hydrokinetic [ˌhaɪdrəʊkaɪˈnetɪk]　　*a.* relating to fluids in motion or the forces that produce or affect such motion 流体动力的;流体动力学的

YSGG laser is a laser-powered hydrokinetic system of new generation, which is now being gradually searched and applied for curing caries.

YSGG 激光是新一代流体动力生物激光系统,在龋病治疗中的应用正不断深入。

obliquely [əˈbliːklɪ]　　*ad.* at an oblique angle 倾斜地

She walked obliquely away across the grass as though trying to get rid of him, and then seemed to resign herself to having him at her side.

她在草地上斜穿过去,好像是要想甩开他,可是后来见到甩不开,就让他走到身旁来。

sporadic [spəˈrædɪk] *a.* recurring in scattered and irregular or unpredictable instances 零星的;分散的

Contrary to official reports, the doctor said that sporadic fighting appeared to be continuing outside the no-fire zone.

与官方报道相反的是,这名医生说,在无战火区之外,零星的战事看似就没有停过。

optimize [ˈɒptɪmaɪz] *v.* to make optimal; get the most out of; use best 使最优化

In this world, the problem and the solution are always the same, and you can effectively optimize the process.

在这一领域,问题与解决方案始终是相同的,而您可以有效地优化该过程。

proponent [prəˈpəʊnənt] *n.* a person who pleads for a cause or propounds an idea 支持者;建议者

He was a major proponent of a health and nutrition program for pregnant women and infants.

他还是孕妇和婴儿健康营养计划的主要支持者。

wane [weɪn] *v.* to become gradually weaker or less 减弱;减少

As the end of January sets in and weight-loss resolutions begin to wane, you may want to try using this technique to boost your efforts.

一月份进入尾声,减肥的决心也开始有些衰退,你可以利用这个技巧来让你事半功倍。

demise [dɪˈmaɪz] *n.* the time when something ends 终止;死亡

Of all the ways we might meet our untimely demise, getting wiped out by an asteroid is the most likely.

我们可能遭遇非命死亡的所有方式中,最有可能的是被小行星彻底消灭。

regulatory [ˈregjʊlətərɪ] *a.* restricting according to rules or principles 管理的;控制的;调整的

For that, I need a regulatory agency that stops households and businesses from polluting the river.

正因为这样,我需要监管机构阻止居民和商业场所污染水质。

intractable [ɪnˈtræktəbl]	*a.* not tractable; difficult to manage or mold 棘手的；难对付的
	Reactive nitrogen emissions from agriculture are the most intractable as they come from many diffuse sources.
	来自农业的活性氮的排放是最棘手的，因为它们的来源很分散。
unduly [ˌʌnˈdjuːlɪ]	*ad.* to an undue degree 过度地；不适当地
	I cannot agree with this school, which seems too individualistic, and unduly indifferent to the importance of knowledge.
	我不同意这所学校的观点，它似乎过于个人主义了，也过分漠视知识的重要性。

➤ *Phrases and Expressions*

in leaps and bounds 飞速地
in particular 尤其；特别
on the horizon 在地平线上；即将来临的
a range of 一系列；一些
the turn of the century 世纪之交
live up to 符合；不辜负（期望）

➤ *Terminologys*

fossil fuels 化石燃料
wave and tidal energy 波浪和潮汐能
ocean thermal energy 海洋热能；海洋温差能
salinity gradient or osmotic energy 盐度梯度或渗透能
offshore wind energy 海上风能；离岸风能
OTEC (Ocean Thermal Energy Conversion) 海洋热能转换

➤ *Proper Names*

UK's Department of Energy 英国能源部
European Marine Energy Centre(EMEC) 欧洲海洋能源中心
Shetland Islands 设得兰群岛
Orkney Islands 奥克尼群岛
R&D (research and development) 研究与开发
House Subcommittee on Oceanography in the United States 美国众议院海洋学小组委员会

➤ *Translation*

1. Paradoxically, the experience gained... devices begin to enter the water. （Para. 1）

矛盾的是，从开采近海化石燃料中获得的经验帮助我们克服了在恶劣的海洋环境中工作的许多技术难题，但随着各种设备开始进入水中，海洋能源正接近一个临界点。

【在翻译的过程中，完全按照原文的句式和字典的释义来进行翻译是行不通的。汉英两种语言在文化背景和语法结构上存在着很大的差异，因此，我们必须根据具体的上下文进行灵活处理。英语中常会出现"overcome the challenges"的表达，如果汉语译成"克服挑战"，就会出现动宾搭配不当的问题，所以要根据上下文译成"克服难题"。】

2. The development of renewable energy technologies... job creation. （Para. 2）

发展可再生能源技术并将其纳入能源组合将减少排放，节约有限的资源并提高能源安全性，此外还将带来额外的共同利益，例如创造就业机会。

【转性译法是英译汉和汉译英中都经常要用到的翻译方法。所谓转性译法就是在翻译过程中，根据译文语言的习惯进行词性转换，如：把原文中的名词转换为动词，把原文中的副词转换为介词，等等。如果把此句的主语完全按照原文词性翻译成"可再生能源技术的发展及其与能源结构的融合"比起前面的译文来就显得不够地道。】

3. Offshore wind in particular... considerable interest and investment. （Para. 2）

特别是海上风能技术，近年来发展迅速，波浪和潮汐技术也开始吸引大量的兴趣和投资。

4. The oscillatory or circular motions... converted into electricity. （Para. 3）

海浪产生的波浪振荡和循环运动可转换成电，而潮水或海洋动能则包括洋流或潮汐，他们可以直接获得并可直接转换成电能。

5. There are floats, flaps, ramps... air-bags and liquid pistons. （Para. 3）

有浮子、挡板、坡道、漏斗、气缸、气囊和液体活塞。

6. In August 2016, the first array... installed capacity of 398 MW. （Para. 6）

2016 年 8 月，第一批潮汐涡轮机向设得兰群岛的电网输送了能量，而 9 月，奥克尼群岛见证了四台 1.5 兆瓦潮汐涡轮机中的第一台下水，该项目的第一阶段最终计划达到 398 兆瓦的装机容量。

7. Devices are advancing rapidly... wave energy projects have been realized. （Para. 6）

设备正在迅速发展，成本正在下降，商业化似乎近在眼前：技术已经得到优化和广泛测试，并且已经实现了先导波能源项目。

8. Despite successful prototype... renewable energy technologies. （Para. 7）

尽管成功完成了原型部署，并开始了首批阵列和大型项目，但海洋能源技术仍然是可再生能源技术中最不发达的。

9. Overall, ocean energy... which concepts would become cost-competitive. （Para. 7）

总体而言，海洋能源技术可能与 20 世纪 70 年代和 80 年代初的风力技术处于同一发展水平，当时正在研究和开发一系列风力涡轮机，但对于哪一种将变得具有成本竞争力仍存在不确定性。

10. While OTEC's demise... potential of ocean energy technologies. （Para. 8）

尽管海洋热能转换技术的衰败在很大程度上是由于石油危机的缓解,但与决策者现在面临的更加棘手的气候和能源问题相比,这仍然提醒人们不要对海洋能源技术的潜力做出过分乐观的评估。

 **Exercises**

1. Fill in the blanks with the proper given words, and then translate the sentences into Chinese.

extent	incredibly	grasp	deployment
exploit	hostile	array	integration
additional	dramatically	potential	quote
extract	funding	phase	risk

1) Even if one day we have an aircraft carrier, unlike another country, we will not use it to pursue global _____ or global reach.

2) In information extraction, it is a common task to _____ concepts such as persons, products, or email addresses from texts.

3) Ford kept his suspicions to himself, even to the _____ of going to jail for a murder he obviously didn't commit.

4) I volunteer because I find something _____ satisfying about acting in a way that is consistent with my principles.

5) I knew we were on to something important when Congress said that if we did this, our _____ would be cut.

6) While you can always delete data, retrieving _____ information can be a painful process once you have lots of it.

7) Not long ago, when his cat died, he cried and understood grief for the first time, but he knows he is difficult to live with because he cannot _____ what others are feeling.

8) The Congresswoman _____ statistics saying that the standard of living of the poorest people had fallen.

9) In the next _____ of the application, the sales representatives can prefill or select information from another domain.

10) Publishers from across the globe have gathered in London this week to discuss how to _____ this growing opportunity.

11) The skipper was not willing to _____ taking his ship through the straits until he could see

where he was going.

12）Many seem to think if they could better express themselves or if their spouse would only listen and understand what they mean then things in the marriage would _____ improve.

13）A lot of times we have to challenge the traditional patterns and make choices that let us use our _____ as well as be ourselves.

14）Drinking may make a person feel relaxed and happy, or it may make her _____, violent, or depressed.

15）These applications do not all have to exist, but for the ones that do, you need to define all the required elements for _____.

16）At the first meeting, last month in Washington, the members listened to an _____ of information and opinions from others.

2. Translate the following passage into Chinese.

Climate change and energy concerns have brought a "cond wave" interest to ocean energy, particularly in the UK, Ireland, Portugal, Denmark, France, Australia, South Korea, Canada, and the United States. Large-scale utilities, energy agencies and industrial companies are making significant investments in the sector, and the European Marine Energy Centre (EMEC), a test bed for emerging ocean energy technologies, is fully subscribed with 14 full-scale devices generating to the grid. Similar testing centres are under development elsewhere. The military has also shown interest in supporting the development of ocean energy technologies, with a US naval base in Hawaii developing a wave energy generation project and a naval base in Western Australia signing an agreement to meet its power needs through ocean energy.

3. Translate the following passage into English.

1608 年,当雨果·格罗蒂乌斯阐述海洋自由原则的时候,广阔而生机盎然的大海似乎确实取之不尽,用之不竭。然而,400 年后,我们的海洋看起来越来越枯竭,人们也越来越意识到我们这个蓝色星球的脆弱性。不断增长的人口和对资源的需求推动了如捕鱼和航运等传统海洋活动的大量增加。同时,一场新的海洋工业革命正在推动对海洋环境的空前开发,这给海洋生态系统带来了进一步的压力,并改变了我们对海洋治理的看法。长期以来,海洋和沿海地区一直是全球经济的强劲驱动力,而沿海社区和港口由于其外向型的地理位置,一直以来都是思想和创新的中心。

Chapter 5

Marine Civil Engineering

Unit 1

Cross-water Civil Engineering

The Golden Gate Bridge

1. The Golden Gate Bridge was once called "the bridge that couldn't be built". Today it is one of the seven wonders of the modern world. This magnificent span, perhaps is San Francisco's most famous landmark, opened in 1937 after a four-year struggle against **relentless** winds, fog, rock and **treacherous** tides.

2. The Golden Gate Bridge is located in San Francisco, USA, on the Golden Gate Strait where San Francisco Bay meets the Pacific Ocean. The Strait itself is approximately 3 miles long by 1 mile wide and is said to have been named by John C. Fremont, a Captain of **Topographical** Engineers in the US Army, around 1846. He chose to call the strait Golden Gate as it reminded him of a harbor in Istanbul known as Chrysoeros that translates to "Golden Horn".

3. Construction of the Golden Gate Bridge started in 1933. The bridge, which was designed by engineer Joseph Strauss, was built to connect San Francisco with Martin County across the 1,600 meters (5,000 feet) wide.

4. The construction of what was to become the world's largest suspension bridge was a **colossal** task. At that time many people did not believe it was technically possible to span the Golden Gate.

5. But despite the disbelief, opposition and the Great Depression, Joseph Strauss was able to find sufficient support and financial backing to go ahead with the project. It would take thousands of workers, four years and 35 million dollars to complete the structure. On May 27, 1937 the Golden Gate Bridge was **inaugurated** by 18,000 people who walked across the bridge. The next day the bridge officially opened to motorized traffic. Today more than 120,000 cars cross the bridge each day.

6. Strauss was chief engineer in charge of overall design and construction of the bridge project. However, because he had little understanding or experience with cable-sus-

pension designs, responsibility for much of the engineering and architecture fell on other experts. Strauss' initial design proposal (two double **cantilever** spans linked by a central suspension segment) was unacceptable from a visual standpoint. The final graceful suspension design was **conceived** and championed by New York's Manhattan Bridge designer Leon Moisseiff.

7. Irving Morrow, a relatively unknown residential architect, designed the overall shape of the bridge towers, the lighting scheme, and Art Deco elements such as the tower decorations, streetlights, railings, and walkways. The famous International Orange color was originally used as a **sealant** for the bridge. The bridge is a lovely shade of "International Orange", or orange **vermilion**, not really gold. It was selected for the way it blends with the natural elements surrounding it. Bridge **lore** tells that the US Navy wanted to paint it black with yellow stripes to be sure it was seen by passing ships. The eye catching orange-red color of the bridge also helped its popularity. The color was suggested by engineer Irving Morrow, who thought the traditional gray color was too boring.

8. Originally the bridge was painted with a lead **primer** and lead based topcoat and this meant only touch up jobs was needed for the next twenty-seven years. In 1968 the maintenance staff realized that the advanced **corrosion** was occurring and so the original paint was removed and replaced with **inorganic** zinc **sulphate** primer and a **vinyl** topcoat. In 1990 the topcoat was again changed to **acrylic emulsion** to meet air quality standards. The maintenance of the paintwork is continual as it gives protection from the high salt content in the air, which can rust or corrode steel components.

9. Moisseiff produced the basic structural design, introducing his "**deflection** theory" by which a thin, flexible roadway would flex in the wind, greatly reducing stress by transmitting forces via suspension cables to the bridge towers. Although the Golden Gate Bridge design has proved sound, a later Moisseiff design, the original Tacoma Narrows Bridge, collapsed in a strong windstorm soon after it was completed, because of an unexpected **aeroelastic** flutter. Ellis was tasked with designing a "bridge within a bridge" in the southern **abutment**, to avoid the need to demolish Fort Point, a pre-Civil War **masonry** fortification viewed, even then, as worthy of historic preservation. He penned a graceful steel arch spanning the fort and carrying the roadway to the bridge's southern **anchorage**.

10. The bridge is popular with pedestrians and bicyclists, and was built with walkways on either side of the six vehicle traffic lanes. Initially, they were separated from the traffic lanes by only a metal curb, but railings between the walkways and the traffic lanes were added in 2003, primarily as a measure to prevent bicyclists from falling in-

to the roadway.

11. One unfortunate situation that affects the importance of the Golden Gate Bridge is the fact that it is the most popular suicide site in the United States. With the deck nearly 250 feet above the **churning** currents, jumpers reach speeds of up to 85 miles an hour by the time they hit the surface of the water. But the impact with the water isn't always fatal — meaning that momentary survivors are dragged by the current over to the rocky **crags** that surround the Golden Gate Strait. The tides act like a washing machine against the **jagged** rocks, making for a violent way to end things.

12. As a prominent American landmark, the Golden Gate Bridge has been used in numerous media, including books, films and video games.

13. The Golden Gate Bridge's significance comes from its rich history as well as the fact that it remained the longest bridge for over 20 years. It still is one of the largest suspension bridges. It was and is the main gateway to the San Francisco Bay. It helped to eliminate the need for ferries, making travel quicker and easier.

14. Many tourists come to San Francisco to see the bridge. Pedestrians and bicyclists often cross the bridge just to enjoy the view and all of the bridge's wonderful characteristics. The bridge's significance will remain throughout its lifetime. The history and hard work that had been put into its creation make it truly outstanding, even if longer suspension bridges are possible.

15. For now, everything remains the same on the Golden Gate Bridge. Vehicles rush to the city and tourists walk along the railings admiring the great view. So if you're one of the lucky ones to visit the landmark, remember where to look and when to get off.

(1,068 words)

> ## *New Words*

relentless [rɪ'lentlɪs]　　　*a.* not stopping or getting less strong 不懈的；不间断的
　　　　　　　　　　　　　　The pressure now was relentless.
　　　　　　　　　　　　　　压力现在没完没了。

treacherous ['tretʃ(ə)rəs]　*a.* dangerously unstable and unpredictable 危险的；变化莫测的
　　　　　　　　　　　　　　The current of the river is fast flowing and treacherous.
　　　　　　　　　　　　　　河水的水流湍急而且变化莫测。

topographical [tɒpə'græfɪk(ə)l]*a.* concerned with topography 地形的；地形学的
　　　　　　　　　　　　　　Due to topographical conditions, truss girder is often used in long-span suspension bridges in the mountain areas.
　　　　　　　　　　　　　　由于受到地形条件所限，山区大跨度悬索桥建设多采用桁架式加劲梁。

colossal [kə'lɒs(ə)l]　　　　*a.* very large 巨大的
　　　　　　　　　　　　　　The colossal monument bestrode the harbor.
　　　　　　　　　　　　　　巨大的纪念碑高耸于海港上。

inaugurate [ɪ'nɔːgjʊreɪt]　　*v.* to commence officially or ceremoniously 正式开始；举行开幕典礼
　　　　　　　　　　　　　　The city library was inaugurated by the mayor.
　　　　　　　　　　　　　　市长主持了市图书馆的落成仪式。

cantilever ['kæntɪliːvə]　　*n.* projecting horizontal beam fixed at one end only 悬臂
　　　　　　　　　　　　　　There is an old steel cantilever bridge over there.
　　　　　　　　　　　　　　那边有一座古老的悬臂钢桥。

conceive [kən'siːv]　　　　*v.* to have the idea for 构思；设想
　　　　　　　　　　　　　　She had conceived the idea of a series of novels, each of which would reveal some aspect of Chinese life.
　　　　　　　　　　　　　　她已经想出了关于一个系列小说的主意，每一部都将反映中国人生活的某一方面。

sealant ['siːlənt]　　　　　*n.* a kind of sealing material that is used to form a hard coating on a porous surface (as a coat of paint or varnish used to size a surface)密封剂
　　　　　　　　　　　　　　The chemical industry says BPA is the safest, most effective sealant.
　　　　　　　　　　　　　　化工行业声称 BPA 是最安全、最有效的密封材料。

vermilion [və'mɪljən]　　　*n.* a variable color that is vivid red but sometimes with an orange tinge 朱红色
　　　　　　　　　　　　　　The furniture on it is glossy vermilion.
　　　　　　　　　　　　　　那上面家具的颜色是闪亮的朱红色。

lore [lɔː]	*n.* knowledge and traditions about a subject or possessed by a particular group of people(某学科的或某部分人的)学问和传统
	bird lore 对鸟类的知识
primer [ˈpraɪmə]	*n.* the first or preliminary coat of paint or size applied to a surface 底漆
corrosion [kəˈrəʊʒ(ə)n]	*n.* erosion by chemical action 腐蚀
	Some substances resist corrosion by air or water. 有些物质可以抵抗空气或水的腐蚀。
inorganic [ɪnɔːˈgænɪk]	*a.* relating or belonging to the class of compounds not having a carbon basis 无机的
	inorganic chemistry 无机化学
sulphate [ˈsʌlfeɪt]	*n.* a salt or ester of sulphuric acid 硫酸盐
	zinc sulphate 硫酸锌
vinyl [ˈvaɪn(ə)l]	*n.* 乙烯基
acrylic [əˈkrɪlɪk]	*a.* 丙烯酸的
emulsion [ɪˈmʌlʃ(ə)n]	*n.* (chemistry) a colloid in which both phases are liquids 乳剂
	acrylic emulsion 丙烯酸乳液
deflection [dɪˈflekʃ(ə)n]	*n.* a twist or aberration 偏向;挠曲
	deflection theory 挠度理论
aeroelastic [eərəʊvɪˈlæstɪk]	*a.* 气动弹性的
abutment [əˈbʌtm(ə)nt]	*n.* structure that bears the weight of a bridge or an arch 桥台
masonry [ˈmeɪs(ə)nrɪ]	*n.* stonework 砖石建筑
	The masonry structure is developing based on brick block. 砖石结构是在砖砌体的基础上发展起来的。
anchorage [ˈæŋk(ə)rɪdʒ]	*n.* place where ships, etc. may anchor safely(船只等安全的)停泊处
churning [ˈtʃɜːnɪŋ]	*a.* (esp. of liquids) moving about violently(尤指液体)翻腾的
	the churning waters of a whirlpool 漩涡状翻腾的水
crag [kræg]	*n.* a steep rugged rock or cliff 峭壁
jagged [ˈdʒægɪd]	*a.* having a sharply uneven surface or outline 参差不齐的
	The stark jagged rocks were silhouetted against the sky. 光秃嶙峋的岩石衬托着天空的背景矗立在那里。

➤ *Phrases and Expressions*

be located in 位于……
remind... of... 使……想起……
go ahead with 继续进行

blend with 与……和谐

touch up 润色;修改

be tasked with 负责

worthy of 值得

➤ *Terminology*

aeroelastic flutter 气动弹性颤振

➤ *Proper Names*

San Francisco 旧金山

Golden Gate Strait 金门海峡

San Francisco Bay 旧金山海湾

John C. Fremont 约翰·C. 弗里蒙特

Istanbul 伊斯坦布尔

Joseph Strauss 约瑟夫·施特劳斯

Marin County 马林郡

the Great Depression 经济大萧条时期

Leon Moisseiff 里昂·摩斯夫

Irving Morrow 欧文·莫罗

Tacoma Narrows Bridge 塔科马纽约湾大桥

➤ *Translation*

1. This magnificent span... opened in 1937 after a four-year struggle against relentless winds, fog, rock and treacherous tides. (Para. 1)

   这座宏伟的桥梁,也许就是旧金山最著名的地标建筑,在经历了不断的风、雾、岩石和变化多端潮汐的四年侵蚀后,终于在 1937 年开放使用。

2. On May 27, 1937 the Golden Gate Bridge was inaugurated by 18,000 people who walked across the bridge. (Para. 5)

   1937 年 5 月 27 日,18 000 人走过了这座桥,宣告了金门大桥的开通。

3. It was selected for the way it blends with the natural elements surrounding it. (Para. 7)

   挑选这种颜色是因为它与金门大桥周围的自然元素很搭配。

4. The maintenance of the paintwork is continual... in the air, which can rust or corrode steel components. (Para. 8)

   漆面的维护工作是持续不断的,以保护桥梁的钢组件不被含盐量较高的空气腐蚀。

5. Moisseiff produced the basic structural design... which a thin, flexible roadway would flex in the wind, greatly reducing stress by transmitting forces via suspension cables to the bridge tow-

ers. (Para. 9)

摩斯夫进行了基本的结构设计,引入了他的"挠度理论",薄而有弹性的路面在风中会弯曲变形,从而通过对桥塔悬挂缆线传输作用力以大大减少压力。

【"挠度理论"来源于稳定理论,属于力学,分为大挠度理论和小挠度理论。主要内容是涉及物体弯曲与力的关系。】

6. Ellis was tasked with designing a "bridge within a bridge" in the southern abutment... a pre-Civil War masonry fortification viewed, even then, as worthy of historic preservation. (Para. 9)

为避免破坏一个建造于内战前但也被看作有历史保护价值的砖石堡垒,艾利斯负责了南桥台的"桥中桥"设计。

【英语中目的状语可以放在一个句子的末尾,但我们将其翻译成汉语的时候,为了符合用语习惯,这个目的状语要先翻译出来放到汉语句子的开头。】

7. Initially, they were separated from the traffic lanes by only a metal curb... and the traffic lanes were added in 2003, primarily as a measure to prevent bicyclists from falling into the roadway. (Para. 10)

起初,人行道和机动车道之间只有金属链来隔离,但2003年又在其间添架了栏杆,来防止非机动车进入机动车道。

8. With the deck nearly 250 feet above the churning currents... an hour by the time they hit the surface of the water. (Para. 11)

高250英尺的甲板下是翻滚的河流,自杀者们击中水面的时速可达85英里。

9. The tides act like a washing machine against the jagged rocks, making for a violent way to end things. (Para. 11)

潮水就像对着锯齿状的岩石翻转的洗衣机一样,使场面往往惨不忍睹。

10. The history and hard work that had been put into its creation make it truly outstanding, even if longer suspension bridges are possible. (Para. 14)

纵使再有更长的吊桥出现,融入这座桥建造过程的历史和辛苦努力都会使这座桥永远傲视群雄。

【我们在翻译限制性定语从句的时候,可以使用前置法,也就是说把英语限制性定语从句译成带"的"的定语词组,放在被修饰词之前,从而将复合句译成汉语单句。】

 **Exercises**

1. Fill in the blanks with the proper given words, and then translate the sentences into Chinese.

relentless	treacherous	colossal	inaugurate
conceive	corrosion	jagged	transmit
preservation	be located in	remind...of...	go ahead with
blend with	touch up	be tasked with	worthy of

1) The message they are _____ to their daughters is very different from that of previous generations.

2) How did they manage to transfer abroad such _____ sums?

3) The weather bureau said it would _____ the installation of a 45 kilometer undersea cable designed to better measure and warn of earthquakes offshore and warn of any potential tsunamis.

4) Men have to be careful what they cry at, because some subjects are more _____ tears than others.

5) He was immensely ambitious but unable to _____ of winning power for himself.

6) The bomb left a pile of _____ glass and twisted metal.

7) Love of not dependable make the person _____ inconstancy quality food easily, hard long-last conservancy.

8) Bitterness is _____ and dangerous when allowed to take root in our hearts.

9) A new corps of European Union guards will _____ patrolling borders.

10) The new shopping mall will _____ the center of town, which can attract more consumers.

11) The surface _____ was worst where the paint had flaked off.

12) Animals' protective coloring enables them to _____ their surroundings.

13) All you need is to spend a little time and effort, giving yourself a little _____.

14) Though so many years have passed by, the paintings were in an excellent state of _____.

15) The current of the river is fast flowing and _____, and it seemed impossible for them to build a bridge over it.

16) If everything goes smoothly, he will be _____ as the president of this country next month.

2. Translate the following passage into Chinese.

Strauss was chief engineer in charge of overall design and construction of the bridge project. However, because he had little understanding or experience with cable-suspension designs, responsibility for much of the engineering and architecture fell on other experts. Strauss' initial design proposal（two double cantilever spans linked by a central suspension segment）was unacceptable from a visual standpoint. The final graceful suspension design was conceived and championed by New York's Manhattan Bridge designer Leon Moisseiff.

3. Translate the following passage into English.

　　第一批桥梁是大自然自己建造的——简单如小溪之上的一根落木或河流当中的石头。第一个由人类制造的桥梁可能是切一段原木或木板甚至石头，做一个简单的支撑和横梁布置。早期美国人利用树木或竹竿来通过小的洞穴或水井，从而从一个地方移动到另一个地方。早期桥梁的常见形式是用长芦苇或其他纤维编织在一起形成一个能够绑定和维系材料的连接绳，把木棍、原木、落叶枝条捆绑在一起。

Unit 2

Underwater Civil Engineering

The Channel Tunnel: the Dream Becomes Reality

1. The idea of constructing a fixed link between France and Britain took root in the imagination of engineers and geologists as early 1751: It has taken nearly 250 years for this dream to be **accomplished**.

2. On 20 January 1986, French President Francois Mitterrand and British Prime Minister Margaret Thatcher announced that the "Transmanche" Project, **proposed** by the Channel Tunnel Group for the U. K. side and the France Manche for the French side, had been selected by both countries.

3. The Franco-British Treaty **concerning** the Channel Tunnel was signed on 12 January 1986 by the Foreign Ministers in Canterbury Cathedral, and the **Concession** Agreement was signed in March 1986.

4. The founder companies, leaders in Public Works in their **respective** countries, set up a Joint Venture under the banner of TML (Transmanche Link).

5. The largest private sector project of the century was given the go-ahead and the works **commenced** at the end of 1986. On 1 December 1990, the historic **junction** was made between France and Great Britain.

6. On 6 May 1994, Queen Elizabeth and President Mitterrand inaugurated the Channel Tunnel.

A few statistics on the Channel Tunnel

7. Below are noted a few **salient** statistics on the Channel Tunnel, chosen from among hundreds that could be listed:
   (1) Eleven Tunnel Boring Machines (TBMs) — five in France and six in the U. K. — were sent into battle on the 150 km of tunnels.
   (2) The project **involved** 554 **transverse** passages; communication passages, piston relief ducts and technical rooms were constructed.
   (3) The data transmission system handles 26,000 items of technical data and 15,000

control points for management of the rail traffic, the **application** software counting 250,000 lines of programme.

(4) The 238-km fibre optic network handles 700 million items of data per second.

(5) Nearly 1,000 partial acceptance tests and 230 system acceptance tests were carried out.

(6) Nearly 13,000 people (5,600 in France) worked directly on the construction of the Channel Tunnel, corresponding to more than 100 million hours of work.

(7) More than 2,000 **subcontractors**, suppliers and consultant practices participated in the construction of the project.

Appropriate logistics

8. From the start of the civil engineering design studies in 1986, the French construction team opted for industrialization and the use of advanced technology in the interest of mastering the tunnel driving programme, as well as to ensure the safety of personnel.

9. In order to achieve these goals, we decided to construct a shaft 55 m in diameter by 65 m deep, situated as close as possible to the French coast.

10. Both tunnels, equipped with two 0.90-m temporary tracks divided into **cantons** 1 km long, were managed by a central traffic control post established in 1989. This temporary post was responsible for all train movements—more than 250 per day.

11. A separate radio channel for each tunnel permitted contact with the locomotives. Each tunnel was managed by a controller, who was provided with a visual control panel showing the path of the programmed **itineraries** and the state of occupation of the cantons.

12. A temporary central electro-mechanical services central post allowed the controller to follow the state of the temporary installations at all times on control monitors. At any moment the controller could modify the equipment settings for electrical supply, **ventilation**, seepage water drainage, gas detection, rail signaling, etc., by remote control. Certain incidents were handled automatically by computers.

13. A 20,000-m^2 segment precasting factory was built a few hundred meters away from the shaft. Between 400 and 500 segments, of 24 different types, were manufactured daily at the factory. The reinforcement cages were manufactured in advance by a three-dimensional **welding** machine. The segments were transported to a stockyard by computer-controlled overhead cranes, to finish curing before being transferred in complete ring sets to the shaft by means of specially designed transporter.

14. These concrete segments have a crushing strength of 70 to 100 newton/mm^2. As a

comparison, the standard for the shell of a nuclear power station is 50 newton/mm^2.

15. These sophisticated techniques allowed us not only to meet the challenge of driving the tunnels, but also to succeed in simultaneously executing the special works (technical rooms, communication passages, crossovers, pumping stations, etc.) and the permanent electro-mechanical installations.

The Channel Tunnel: a human gamble

16. The success of the Channel Tunnel depended on meeting a double challenge:
    (1) Implementing and mastering the exceptional techniques required for the works; and
    (2) maintaining good labor relations.

17. From the start of recruiting personnel, the management of French construction decided to carry out the works using local personnel to the maximum, since the rate of unemployment in the "Nord-Pas-de-Calais" region had reached 22%.

18. In order to do this, a unique training and tutorial programme was established with the help of the national and regional government authorities. The programme involved providing 183,000 hours of site access training and 560,000 hours of further on-the-job training so that people who initially had little experience and training in our profession could be qualified to work on the project.

19. The results speak for themselves: 95% of the manual labor and 68% of the management and **supervisory** staff were recruited in the region.

20. Within the training function, safety occupied an important place. Five hundred senior staff attended a two-day safety conference; and 700 individuals were trained in emergency and first aid procedures.

21. At the start of method studies for the execution of the structural works and the electro-mechanical installations, a risk analysis was undertaken. In addition, protective safety measures for incorporation in the machines were taken into account right at design stage, and were checked in partnership with safety organizations. More than 300 studies of this type were carried out.

22. Responsibility was delegated through the management chain in all fields with respect to safety, quality, cost and programme. These measures permitted reduction of the work accident rate to 50% that of the national average for our industry; and created an exceptional climate of labor relations.

(1,002 words)

➤ *New Words*

accomplish [əˈkʌmplɪʃ]

    v. to put in effect 完成;实现

    If we'd all work together, I think we could accomplish our goal.

    如果我们齐心协力,我想我们能实现我们的目标。

propose [prəˈpəuz]

    v. to put forward 提议;提出

    The resolution was proposed by the chairman of the International Committee.

    那项决议是由国际委员会主席提出的。

concerning [kənˈsɜːnɪŋ]

    prep. about or relating to 关于,有关,涉及

    He heard nothing concerning this matter.

    关于这件事他什么都没听到。

concession [kənˈseʃ(ə)n]

    n. a contract granting the right to operate a subsidiary business 特许

    He got the beer concession at the ball park.

    他得到了在棒球场喝啤酒的特许。

respective [rɪˈspektɪv]

    a. considered individually 分别的;各自的

    We all went back to our respective homes to wait for news.

    我们都各自回家等待消息。

commence [kəˈmens]

    v. to begin or to start something 开始;着手

    Work will commence on the new building immediately.

    新大楼即将破土动工。

junction [ˈdʒʌŋkʃ(ə)n]

    n. the state of being joined together 连接;接合

    The cars collided at the junction.

    汽车在交叉路口相撞。

salient [ˈseɪlɪənt]

    a. having a quality that thrusts itself into attention 显著的;突出的

    Chronic fatigue is also one of the salient features of depression.

    慢性疲劳也是抑郁症的显著特点之一。

involve [ɪnˈvɒlv]

    v. to include 包含;涉及

    However, the use of these materials does involve some inherent questions.

    然而,对这些材料的使用涉及一些固有的问题。

transverse [ˈtrænzvɜːs]

    a. extending or lying across 横的;横向的

    Machine allows transverse left/right moving, adjusted by

electric motor.

本机器可左、右横移,以电动机方式调整。

application [ˌæplɪˈkeɪʃ(ə)n]　　*n.* using it for a particular purpose 应用

How do the two techniques compare in terms of application?

这两种手法实际运用起来哪个好一些?

subcontractor [ˈsʌbkəntræktə(r)]　*n.* someone who enters into a subcontract with the primary contractor 转包商

The company was considered as a possible subcontractor to build the aeroplane.

该公司被视为能够承担该飞机制造任务的潜在分包商之一。

logistics [ləˈdʒɪstɪks]　　*n.* handling an operation that involves providing labor and materials be supplied as needed 后勤

These are the logistics of what I have to do to get work in this field.

我必须做这些筹备才在这个领域能找到工作。

canton [ˈkæntɔn]　　*n.* a small administrative division of a country 行政区

the Swiss canton of Berne

瑞士伯尔尼地区

itinerary [aɪˈtɪn(ə)r(ə)rɪ]　　*n.* an established line of travel or access 旅程;路线

The next place on our itinerary was Silistra.

我们行程的下一站是锡利斯特拉。

ventilation [ˌventɪˈleɪʃ(ə)n]　　*n.* a mechanical system in a building that provides fresh air 通风设备

The only ventilation comes from tiny sliding windows.

只能通过小推拉窗来通风。

welding [ˈweldɪŋ]　　*n.* fastening two pieces of metal together by softening with heat and applying pressure 焊接

All the welding had been done from inside the car.

所有的焊接都在车内完成。

supervisory [ˈsjuːpəˌvaɪzərɪ]　　*a.* of or limited to or involving supervision 监督的

Most supervisory boards meet only twice a year.

大多数监事会一年只开两次会议。

➤ *Phrases and Expressions*

take root in 根植于;孕育于
corresponding to 与……一致;相当于
participate in 参加;参与
in the interest of 为了……的利益
be responsible for 为了
take... into account 把……考虑在内

➤ *Terminology*

Public Works 公共工程
Joint Venture 合资企业
Tunnel Boring Machines 隧道掘进机
piston relief ducts 活塞调剂管道
seepage water 渗透水
precasting factory 预制构件厂

➤ *Proper Names*

Francois Mitterrand 弗朗索瓦·密特朗
Margaret Thatcher 玛格利特·撒切尔
"Transmanche" Project 跨芒什海峡工程
Manche 芒什省(法国省份)
Canterbury Cathedral 坎特伯雷大教堂
Queen Elizabeth 伊丽莎白女王

➤ *Translation*

1. The largest private sector project of the century was given the go-ahead and the works commenced at the end of 1986. (Para. 5)

   20 世纪最大的私营工程于 1986 年底开始了。

   【无论是英译汉,还是汉译英,有时为了行文的需要,都可以在行文里增加几个词,或减少几个词。增词可以使译文流畅,减词可以使译文简洁。这是因为有些词在一种语言里可能是必要的,而在另一种语言里就会显得多余了。】

2. The data transmission system handles 26,000 items of technical data and 15,000... the application software counting 250,000 lines of programme. (Para. 7)

   数据传输系统要处理 26 000 项技术数据和 15 000 个铁路交通管理控制点,应用软件要计算 250 000 条程序。

3. From the start of the civil engineering design studies in 1986... and the use of advanced technology in the interest of mastering the tunnel driving programme, as well as to ensure the safety of personnel. (Para. 8)

从 1986 年土木工程的设计研究开始,法国建筑队已经产业化了,他们用先进的技术去控制隧道的钻进程序,也确保了人员的安全。

4. Both tunnels... into cantons 1 km long, were managed by a central traffic control post established in 1989. (Para. 10)

两条隧道各装备有两条 0.9 米宽 1 千米长的临时轨道,由一个建于 1989 年的中央交通控制岗位管理。

5. Each tunnel was managed... with a visual control panel showing the path of the programmed itineraries and the state of occupation of the cantons. (Para. 11)

每条隧道都由一个管理员管理,这个管理员操作一个可视化屏幕,它显示了编制的路线和各分区的占用情形。

【如果英语的定语从句结构复杂,译成汉语前置定语显得太长而不符合汉语表达习惯时,往往可以译成后置的并列分句。】

6. A temporary central electro-mechanical services central post allowed... the temporary installations at all times on control monitors. (Para. 12)

一个临时的中央电力机械服务中心岗位允许控制员在监视器上随时跟踪临时设备的进程。

7. The segments were transported to a stockyard by computer-controlled overhead cranes... in complete ring sets to the shaft by means of specially designed transporter. (Para. 13)

构件由计算机控制的高空起重机送到堆场,以便再完成养护,最后又经过特殊设计的运输机全部运送到管道中。

8. From the start of recruiting personnel... decided to carry out the works using local personnel to the maximum, since the rate of unemployment in the "Nord-Pas-de-Calais" region had reached 22%. (Para. 17)

从开始征募工作人员时,法国建造部分的管理层就决定尽可能地使用当地人员,因为加来北部地区的失业率已经达到 22%。

9. The programme involved providing 183,000 hours of site access training and 560,000 hours... so that people who initially had little experience and training in our profession could be qualified to work on the project. (Para. 18)

这套程序包括提供 183 000 小时的现场训练和 560 000 小时更深入的实际操作训练,这样就使得本来没有本行业经验和训练的工人最终能够符合工程的需要。

10. In addition, protective safety measures for incorporation in the machines were taken into account right at design stage... with safety organizations. (Para. 21)

此外,在设计阶段也考虑了结合机器设备的防范措施,并与安全组织一同进行考核。

11. These measures permitted reduction of the work accident rate to 50%... and created an ex-

ceptional climate of labor relations. (Para. 22)

这些措施能使工作中的意外事件发生概率降到法国国内同行业平均值的 50%，并创造了良好的工作关系。

 **Exercises**

1. Fill in the blanks with the proper given words, and then translate the sentences into Chinese.

accomplish	propose	concerning	respective
corresponding to	commence	involve	carry out
relief	application	be responsible for	itinerary
take... into account	supervisory	in the interest of	

1) Students learned the practical _____ of the theory they had learned in the classroom.

2) Newton _____ that heavenly and terrestrial motion could be unified with the idea of gravity.

3) Especially the effect on _____ the policy to serve agriculture, countryside and farmer has an incomparable advantage over other rural financial organizations.

4) Nicky's job as a public relations director _____ spending quite a lot of time with other people.

5) Such trends in my field and yours have implications for the identity of our _____ disciplines.

6) The calculations do not _____ any fluctuation in the share price.

7) I'm ready to accept any job whatever, so long as it is _____ the people.

8) The engineer and his son held frequent consultations _____ technical problems.

9) To _____ his own end, he placed collective interests in the back of his mind.

10) _____ agencies are stepping up efforts to provide food, shelter and agricultural equipment.

11) The administrative department of economy and trade under the State Council shall _____ organizing and conciliating the promotion of clean production within the whole country.

12) This future is not fixed, but it is a destination that can be reached if we pursue a sustained dialogue like the one that you will _____ today, and act on what we hear and what we learn.

13) Appraisal has traditionally been seen as most applicable to those in management and _____ positions.

14) These conditions may be regarded as _____ a shallow underwater explosion.

15) If you're running a travel site, or a bank, you can use that pipeline to transmit structured data to your users—for example, _____ or transaction reports.

2. Translate the following passage into Chinese.

A 20,000 m^2 segment precasting factory was built a few hundred meters away from the shaft. Between 400 and 500 segments, of 24 different types, were manufactured daily at the factory. The reinforcement cages were manufactured in advance by a three-dimensional welding machine. The segments were transported to a stockyard by computer-controlled overhead cranes, to finish curing before being transferred in complete ring sets to the shaft by means of specially designed transporter.

3. Translate the following passage into English.

英吉利海峡隧道虽历经两个世纪才建成,但在 1994 年 5 月 6 日,这个耗时 6 年的项目成为英国和欧洲大陆之间首条坚实的运输线。从那以后,这条隧道既可民用又可商用,虽耗资有些大,但速度相对较快。虽然已证实它并未能取得预期的经济效益,但这条隧道仍是工程领域中前所未有的奇迹,并在 1996 年被美国土木工程师协会评为现代世界七大奇观之一。

Unit 3
Coastal Civil Engineering

Shoring up Coastal Engineering

1. Coastal engineering is a relatively new field. It has arisen from the need for specialized training in the **complexities** of coastal processes and in the design and construction techniques suited to the coastal environment. Unfortunately, the number of coastal engineers is not keeping pace with the growing need for their skills, and unless this area receives attention at the national level the future will not be particularly bright for coastal engineers or for coastal communities.

2. The greatest challenge facing the discipline of coastal engineering and one of the many strong arguments for strengthening the field—is that the number of people living and working on or near US coastlines is growing. Half the population of the United States now lives within 50 miles (80 km) of a coast, according to the National Oceanic and **Atmospheric** Administration (NOAA). Aside from the fact that many people find coastal living appealing and the nation's shorelines attract more tourists than do national parks, the coastlines are becoming more important to the nation's economic well-being. According to NOAA, one in six jobs in the United States is now marine related, and the **anticipated** tripling of imported cargo by 2020 will require expanded harbor facilities and deeper **navigational** channels to **accommodate** the newer cargo ships.

3. At the same time, however, coastal hazards, changes in climate, extensive shoreline erosion, and costly port maintenance are combining to make coastal areas both dangerous and expensive. In fact, the portion of coastal **infrastructure** vulnerable to coastal erosion along the East and Gulf coasts along is valued at $3 trillion, and coastal property with an insured value of $2 trillion lies within zones affected by At-

lantic hurricanes.

4. These are among the major research topics that must be addressed to better understand the processes occurring in the surf zone and to better estimate the performance of structures or other human interventions in the coastal zone. Basic research provides the scientific and physical bases for many of the coastal engineer's **predictive** tools. The near-shore zone is an extremely complicated region, and **sediment** transport on beaches is a notoriously difficult and unsolved problem. Without more research on an accurate way to predict the amount of material moving along a shoreline, engineering designs must rely even more on historical records and the skill of the coastal engineer.

5. Coastal processes in the **vicinity** of tidal entrances are not well understood and cannot be accurately predicted. Neither is shoaling of navigational channels fully predictable, and waves in the channels and harbors need to be better modeled to help solve certain navigation problems. Data that should be obtained and maintained on the coastal environment include long-term wave measurements and **periodic** beach surveys, which are needed to provide design variables.

6. Applied research could aid the development of coastal models to predict the behavior of the shoreline over short periods of time, such as during a storm, as well as over long periods to see the effects of, for example, the construction of a **jetty** at an inlet. Coastal simulators that provide a realistic picture of the surf zone would be of great use to the military for the planning of coastal operations.

7. Few of these programs are associated with major research laboratories where physical modeling of coastal processes and waves may be conducted, and not all of the laboratories are state of the art—the result of insufficient funding. Yet physical modeling is a very beneficial tool in the discovery and **elucidation** of important processes. Significant findings at the University of Delaware during the past three years have included the discovery of instabilities in rip currents as they flow seaward, causing them to **oscillate** from side to side and to **pulsate** in strength; the **verification** of the instability of the undertow as it flows offshore, leading to a theory to explain the presence of large horizontal **eddies** lying along the breaker line; and the nonlinear behavior of waves as they are blocked by ebb currents in the vicinity of inlets.

8. First, coastal engineering professors need to establish an academic **consortium** to improve research and educational opportunities and to provide educational leadership for

curricular issues and for educating the general public about the field. Second, the National Science Foundation (NSF) should form and fund a focused program for coastal engineering research. This recommendation was first made in a 1984 NSF report (Natural Hazards and Research Needs in Coastal and Ocean Engineering). At that time it was recommended, without effect, that $10 million be dedicated annually to study the effects of such phenomena as hurricanes, sea level rise, **tsunamis**, beach erosion, and harbor **siltation**, but the same types of problems remain more than 15 years later. And third, the US Army Corps of Engineers should fund basic research in academic institutions. Although the Corps conducts applied research in-house at its Coastal and **Hydraulics** Laboratory, it does not conduct basic research. It needs to develop a better relationship with universities so that basic research—is carried out. There are examples of fruitful **collaboration** between industry and **academia** in Europe—particularly in Denmark and Holland—that may serve as models.

9. During the past year the coastal engineering profession has responded to the Marine Board's report by founding the Association of Coastal Engineers, in Alexandria, Virginia, and ASCE has established the Coasts, Oceans, Ports, and Rivers Institute. In addition, the National Science Foundation has just funded a tsunami wave basin in Corvallis at Oregon State University, that promises to be an important resource for the coastal engineering community. While these are positive steps, much more needs to be done to ensure that the nation has a sufficient supply of well-trained coastal engineers in the future.

(955 words)

➢ *New Words*

complexity [kəmˈpleksətɪ]
 n. the quality of being intricate and compounded 复杂性
 The situation presents great complexity.
 形势显得极为复杂。

atmospheric [ˌætməsˈferɪk]
 a. relating to or located in the atmosphere 大气的, 大气层的
 The barometer marked a continuing fall in atmospheric pressure.
 气压表表明气压在继续下降。

anticipate [ænˈtɪsəˌpeɪt]
 v. to expect sth. 预料; 预期
 We anticipated her winning first prize.
 我们期望她获得第一名。

navigational [ˌnævɪˈgeɪʃənəl]
 a. of or relating to navigation 航行的, 航运的
 Gyroscopes are used in navigational instruments for ships, aircraft, and spacecraft.
 陀螺仪用于船舶、飞机和宇宙飞船中的导航设备。

accommodate [əˈkɒmədeɪt]
 v. to be agreeable or acceptable to 容纳; 使适应
 She tried to accommodate her way of life to his.
 她试图使自己的生活方式与他的生活方式相适应。

infrastructure [ˈɪnfrəstrʌktʃə]
 n. the basic structure or features of a system or organization 基础设施
 The infrastructure, from hotels to transport, is old and decrepit.
 从宾馆到交通, 所有的基础设施都已陈旧失修。

predictive [prɪˈdɪktɪv]
 a. of or relating to prediction 预言性的; 成为前兆的
 They acquire authority from their predictive power.
 他们的权威性来自其预测事物的能力。

sediment [ˈsedɪm(ə)nt]
 n. matter that has been deposited by some natural process 沉积; 沉淀物
 The underground sediment of the Ganges Delta contains arsenic.
 恒河三角洲的地下沉积物中含有砷。

vicinity [vɪˈsɪnɪtɪ]
 n. a surrounding or nearby region 邻近, 附近; 近处
 He told us there was no hotel in the vicinity.
 他告诉我们说附近没有旅馆。

periodic [ˌpɪərɪˈɒdɪk]
 a. happening or recurring at regular intervals 周期的; 定期的

Periodic checks are carried out on the equipment.

设备定期进行检查。

jetty ['dʒetɪ] 　　　　*n.* a protective structure of stone or concrete; extends from shore into the water to prevent a beach from washing away 码头;防波堤

The boat was tied up alongside a crumbling limestone jetty.

这条船停泊在一个摇摇欲坠的石灰岩码头边。

elucidation [ɪˌluːsɪ'deɪʃ(ə)n] 　　　　*n.* an interpretation that removes obstacles to understanding 说明;阐明

It is a straight-forward, scientific, clear elucidation of facts that cannot be disputed.

这些是直接的、科学的、明确的、不能争议的事实阐释。

oscillate ['ɒsɪleɪt] 　　　　*v.* to move or swing from side to side regularly 振动;振荡;摆动

The price of grain slightly oscillated last year.

去年的谷物价格略有波动。

pulsate [pʌlˌseɪt] 　　　　*v.* to move with or as if with a regular alternating motion 搏动;悸动;有规律地跳动

The air seemed to pulsate with the bright light.

空气似乎随着亮光而颤动。

verification [ˌverɪfɪ'keɪʃ(ə)n] 　　　　*n.* additional proof that something that was believed (some fact or hypothesis or theory) is correct 确认,查证;核实

All charges against her are dropped pending the verification of her story.

在她的事情核实之前,所有针对她的指控都已撤回。

eddy ['edɪ] 　　　　*n.* a miniature whirlpool or whirlwind resulting when the current of a fluid doubles back on itself 涡流;漩涡;逆流

The motor car disappeared in eddy of dust.

汽车在一片扬尘的涡流中不见了。

consortium [kən'sɔːtɪəm] 　　　　*n.* an association of companies for some definite purpose 社团,财团;组合

The consortium includes some of the biggest building contractors in Britain.

该联合企业包括英国最大的一些建筑承包商。

tsunami [tsuː'nɑːmɪ] 　　　　*n.* a cataclysm resulting from a destructive sea wave caused by an earthquake or volcanic eruption 海啸

Powerful quake sparks tsunami warning in Japan.

大地震触发了日本的海啸预警。

siltation [ˈsɪlɪˌteɪʃən]   *n.* the pollution of water by fine particulate terrestrial clastic material, with a particle size dominated by silt or clay 淤积;聚积

The siltation relates with the circumfluence.

淤积的主要原因是回流淤积。

hydraulics [haɪˈdrɔːlɪks]   *n.* study of the mechanics of fluids 水力学

Hydraulics Laboratory

水力学实验室

collaboration [kəˌlæbəˈreɪʃn]   *n.* act of working jointly 合作

There is substantial collaboration with neighboring departments.

与相邻的一些部门有大量的合作。

academia [ˌækəˈdiːmɪə]   *n.* the academic world 学术界

The government and academia should pay attention to this unusual fact.

这种不寻常现象应引起政府监管部门与学术界的关注。

➤ *Phrases and Expressions*

arise from 起因于

keep pace with 保持同步

aside from 此外

state of the art 最新水平;最高水平

be dedicated to 致力于;从事于

➤ *Terminology*

rip currents 裂口流

ebb currents 落潮流

➤ *Proper Names*

National Oceanic and Atmospheric Administration (NOAA) 国家海洋和大气管理局

the University of Delaware 特拉华大学

the National Science Foundation (NSF) 国家科学基金会

the US Army Corps of Engineers 美国陆军工程兵团

Denmark 丹麦

Holland 荷兰

the Marine Board 海洋局

Alexandria 亚历山大市

Corvallis 科瓦利斯

▶ *Translation*

1. It has arisen from the need for specialized training in the complexities of... （Para. 1）

   它的兴起源于复杂的海岸处理技术和与海岸环境相适应的设计施工技术所需要的专门训练。

2. According to NOAA, one in six jobs in the United States is now marine related, and the anticipated tripling of imported cargo... （Para. 2）

   根据国家海洋和大气管理局统计,美国有六分之一的工作与海洋有关,预计到 2020 年进口货物将增加到现在的三倍,这就要求扩展海港设备和加深航道以容纳更新的货轮。

   【在英译汉过程中,有些句子可以逐词对译,有些句子则由于英汉两种语言的表达方式不同,就不能用"一个萝卜一个坑"的方法来逐词对译。原文中有些词在译文中需要转换词类,才能使汉语译文通顺自然。比如此句翻译中,我们就可以把形容词 expanded 和 deeper 转换成动词意义,译文会变得更通顺。】

3. In fact, the portion of coastal infrastructure vulnerable to coastal erosion along the East and Gulf coasts along is valued at ＄3 trillion, and coastal property with an insured value... （Para. 3）

   实际上,仅仅沿着东部和海湾海岸线,易受沿海腐蚀的沿海基础设施部分的价值就达 3 万亿美元,并且有价值 2 万亿美元投保金额的沿海财产位于大西洋飓风侵袭的区域内。

   【英汉两种语言都有一词多类、一词多义的现象。一词多类就是指一个词往往属于几个词类,具有几个不同的意义。一词多义就是指同一个词在同一个词类中,又往往有几个不同的意义。在英汉翻译过程中,我们在弄清原句结构后就要善于选择和确定原句中关键词的词义。比如,我们可以根据词在句中的词类来选择和确定词义。本句中 along 出现了两次,它们的词类是不同的,我们可以根据词类来分别确定它们各自的意义,第一个 along 为介词,第二个 along 为副词。】

4. These are among the major research topics that must be addressed to better understand the processes occurring in the surf zone and to better estimate... （Para. 4）

   这些就是主要的研究课题,以更好地了解发生在碎浪地区的过程和更好地估计结构性能或人类对海岸地区的其他干预。

5. Without more research on an accurate way to predict the amount of material moving along a shoreline... the coastal engineer. （Para. 4）

   在预测沿着海岸线移动的物质的数量上,若没有对精确方法的进一步研究,工程的设计就更加依赖于历史记录和海岸工程师的技能。

6. Neither is shoaling of navigational channels fully predictable... and harbors need to be better modeled to help solve certain navigation problems. （Para. 5）

   航海通道中的浅滩化不可能被完全预测,通道和港湾中的波浪需要更好地建模以便有助于解决某些航行问题。

7. Applied research could aid the development of coastal models to predict the behavior of the shoreline over short periods of time, such as during a storm... the construction of a jetty at an inlet. (Para. 6)

应用研究能帮助沿海模型的发展,从而能在短期内预测海岸线的状态,比如在暴风雨期间或者是在长时间内,人们都可以观察,例如,在入口处建设码头所带来的影响。

8. Significant findings at the University of Delaware during the past three years have included the discovery of instabilities in rip currents as they flow seaward... and the nonlinear behavior of waves as they are blocked by ebb currents in the vicinity of inlets. (Para. 7)

在过去三年中,特拉华大学重要的发现包括:当流向海洋时裂口流是非稳定的,这导致了他们从一侧到另一侧的摆动及能量上的波动;下层逆流在离岸方向流动时是不稳定的,这引出了沿破坏线会存在大型水平漩涡的理论;以及在入海口附近波浪被退潮阻挡时的非线性特征。

9. First, coastal engineering professors need to establish an academic consortium... for curricular issues and for educating the general public about the field. (Para. 8)

首先,海岸工程师需要建立一个学术协会来促进研究和增加教育机会,并在解决课程问题和向大众普及该领域的知识方面发挥其教育领导力的作用。

10. In addition, the National Science Foundation... that promises to be an important resource for the coastal engineering community. (Para. 9)

另外,国家科学基金会刚刚在科瓦利斯的俄勒冈州立大学资助了海啸波区池,它必将成为海岸工程学领域一个重要的资源。

 **Exercises**

1. Fill in the blanks with the proper given words, and then translate the sentences into Chinese.

complexity	atmospheric	anticipated	navigational
accommodate	infrastructure	predictive	periodic
elucidation	verification	collaboration	arise from
keep pace with	aside from	be dedicated to	

1) If you happen to find this approach successful, please let us know so that we can plan _____ updates for future columns.

2) _____ that and the policy on our two cars at ＄400 a year, we have no other insurance.

3) They are simply an indication of _____ implementations that may or may not be true when we complete the service specifications and implementations.

4) More complex solutions also increase the _____ of planning, development, and manufacturing processes.

5) Scientists hope the work done in _____ with other researchers may be duplicated elsewhere.

6) His mother had been told by an angel that she would bear a son whose life would _____ God and whose hair must never be cut.

7) Some animal and plant species cannot _____ to the rapidly changing conditions.

8) This allowed initial system testing and problem determination to be performed very shortly after functional _____ completed.

9) The energy from these _____ waves, like the energy from a sound wave, propagates both horizontally and vertically.

10) The subsequent isolation and _____ of the active compound by laboratory scientists, they argue, can be relatively routine tasks.

11) Lead contamination of food can also _____ food processing, food handling, and food packaging.

12) _____ construction is a significant part of Africa's further economic and social development.

13) We will see again a consumer that can _____ the economy, but cannot drive the economy forward.

14) _____ control is a new type of computer control algorithms being widely noticed and used in industrial process control.

15）For long-distance trips or when they were over unfamiliar territory, pigeons use their own _____ system.

2. Translate the following passage into Chinese.

These are among the major research topics that must be addressed to better understand the processes occurring in the surf zone and to better estimate the performance of structures or other human interventions in the coastal zone. Basic research provides the scientific and physical bases for many of the coastal engineer's predictive tools. The near-shore zone is an extremely complicated region, and sediment transport on beaches is a notoriously difficult and unsolved problem. Without more research on an accurate way to predict the amount of material moving along a shoreline, engineering designs must rely even more on historical records and the skill of the coastal engineer.

3. Translate the following passage into English.

由于陆地面积的不足,土木工程师开始想办法去利用地球上的其他空间,例如海洋。所以,填海便成为缓解这一现状的方法。这确实可以为一些国家增加陆地面积。更多的楼群和基础设施可以建在开发出来的陆地上。在新加坡,填海活动正在紧锣密鼓地进行着,通过填海获得的陆地随处可见,例如举世闻名的新加坡樟宜机场,就建造在这样的地面之上。包括公园、人造海滩和住宅的整个东海岸公园也建造在新加坡最大的新生地上。在临近岛屿上,一些居住用、工业用及其他公用的建筑也坐落在这类地面上。

Chapter 6

Marine Transportation

Unit 1
Marine Geography and Traffic

Marine Geography and Routes

1. Marine transportation concerns the movement of passengers and freight over water masses, from oceans to rivers. From its modest origins as Egyptian coastal and river sailships around 3,200 BCE, marine transportation has always been the dominant support of global trade. By 1,200 BCE Egyptian ships traded as far as Sumatra, representing one of the longest marine routes of that time. By the 10th century, Chinese merchants frequented the South China Sea and the Indian Ocean, establishing regional trade networks. In the early 15th century, Admiral Zheng He led a large fleet of more than 300 vessels manned by a crew of 28,000 to conduct seven major expeditions, one of which reached the East African coast. However, China's attempt at asserting a regional marine dominance was short-lived and such expeditions were not permitted to continue mostly because China perceived itself as a continental power with marine trade of limited interest.

2. However, for other nations, the projection of marine power became of strategic interest. European colonial powers, mainly Spain, Portugal, England, the Netherlands, and France, would be the first to establish a true global marine trade network from the 16th century. Most of the marine shipping activity focused around the Mediterranean, the northern Indian Ocean, Pacific Asia and the North Atlantic, including the Caribbean. Thus, access to trade commodities remains historically and contemporarily the main driver in the setting of marine networks.

3. With the development of the steam engine in the mid-19th century, trade networks expanded considerably as ships were no longer subject to dominant wind patterns. Accordingly, and in conjunction with the opening of the Suez Canal, the second half of the 19th century saw an **intensification** of marine trade to and across the Pacific. In the 20th century, marine transport grew **exponentially** as changes in international trade and seaborne trade became interrelated. Marine transportation, like all transportation, is a derived demand that exists to support trade relations. These trade rela-

tions are also influenced by the existing marine shipping capacity and the changes in the composition of marine shipping services. There is thus a level of **reciprocity** between trade and marine shipping capabilities. As of 2008, seaborne trade accounted for 89.6% of global trade in terms of volume and 70.1% in terms of value. Marine shipping is one of the most globalized industries in terms of ownership and operations.

4. Marine transportation, like land and air modes, operates on its own space, which is at the same time geographical by its physical **attributes**, strategic by its control and commercial by its usage. While geographical considerations tend to be constant in time (except for the seasonality of weather patterns), strategic and especially commercial considerations are much more dynamic. The **physiography** of marine transportation is composed of two major elements, which are rivers and oceans. Although they are connected, each represents a specific domain of marine circulation. The notion of marine transportation rests on the existence of regular **itineraries**, better known as marine routes.

5. Marine routes are corridors of a few kilometers in width connecting economic regions and overcoming the **discontinuities** of land transport. They are a function of obligatory points of passage, which are strategic places, of physical constraints (coasts, winds, marine currents, depth, reefs, ice) and of political borders. As a result, marine routes draw arcs on the earth water surface as intercontinental marine transportation tries to follow the great circle distance. Marine routes are linking marine ranges representing main commercial areas between and within which marine shipping services are established.

6. The most recent technological transformations affecting water transport have focused on modifying water channels (such as **dredging** port channels to deeper depths), on increasing the size, the automation and the specialization of vessels (e.g. container ships, tanker, bulk carrier) and developing massive port terminal facilities to support the technical requirements of marine transportation. These transformations partially explain the development of marine traffic that has been adapting to increasing energy demand (mainly fossil fuels), the movements of raw materials, the location of major grain markets and to the growth of the trade of intermediate and finished goods. Yet, this process is not uniform and various levels of connectivity to global shipping networks are being observed. The **massification** of transport into regular flows over long distances is not without consequences when accidents affecting oil tankers can lead to major ecological disasters (e.g. Amoco Cadiz, Exxon Valdez).

7. **Fluvial** transportation, even if slow and inflexible, offers a high capacity and a continuous flow. The fluvial / land **interface** often relies less on transshipment **infra-**

structures and is thus more permissive for the location of dependent activities. Ports are less relevant to fluvial transportation but fluvial hub centers experience a growing integration with marine and land transportation, notably with containerization. The degree of integration for fluvial transportation varies from totally isolated distribution systems to well-integrated ones. In regions well supplied by **hydrographic** networks, fluvial transportation can be a privileged mode of shipment between economic activities. In fact, several industrial regions have emerged along major fluvial axis as this mode was initially an important vector of industrialization. More recently, river-sea navigation is also providing a new dimension to fluvial transportation by establishing a direct interface between fluvial and marine systems.

8. Most marine circulation takes place along coastlines and three continents have limited fluvial trade; Africa, Australia, and Asia (with the exception of China). There are however large fluvial waterway systems in North America, Europe, and China over which significant fluvial circulation takes place. Fluvial-marine ships are able to go directly from fluvial to oceanic marine networks. Despite regular services on selected fluvial **arteries**, such as the Yangtze, the potential of waterways for passenger transport remains limited to fluvial tourism (river cruises). Most major marine infrastructures involve maintaining or modifying waterways to establish more direct routes (navigation channels and canals). This strategy is however very expensive and undertaken only when absolutely necessary. Significant investments have been made in expanding transshipment capacities of ports, which is also very expensive as ports are heavy consumers of space.

9. Not every region has direct access to the ocean and marine transport. As opposed to coastal countries, maritime **enclaves** (landlocked countries) are such countries that have difficulties to undertake marine trade since they are not directly part of an oceanic domain of marine circulation. This requires agreements with neighboring countries to have access to a port facility through a highway, a rail line or through a river. However, being landlocked does not necessarily imply exclusion from international trade, but substantially higher transport costs which may impair economic development. Further, the concept of being landlocked can be relative since a coastal country could be considered as relatively landlocked if its port infrastructures were not sufficient to handle its marine trade or if its importers or exporters were using a port in a third country. For instance, France has significant **nautical** accessibility, but the main port handling its containerized traffic is Antwerp in Belgium.

10. The importance and **configuration** of marine routes have changed with economic development and technical improvements. Among those, containerization changed the configuration of freight routes with innovative services. Prior to containerization,

loading or unloading a ship was a very expensive and time-consuming task and a cargo ship typically spent more time docked than at sea. While sailing time used to represent around 25% of the annual ship time for standard break-bulk ships, this figure is now around 70% for containerships. With faster and cheaper port operations, inter-range routes have emerged as a dominant configuration of containerized marine networks.

11.   Inter-range service involves a set of sequential port calls from at least two marine ranges, commonly including a transoceanic service and structured as a continuous loop. They are almost exclusively used for container transportation with the purpose of servicing a market by balancing the number of port calls and the frequency of services.

12.   The main advantage of inter-range services is the ability to call several ports and therefore increase the ship load factor. This sequence of ports tends to be highly flexible in terms of which ports are serviced to maximize the market potential. There is however the risk of empty trips (particularly **backhauls**) and longer service times between distant port pairs along the route. The first inter-range route was set in 1962 by Sea-Land between the ports of New York (Newark facilities), Los Angeles and Oakland by using the Panama Canal. The return trip also included a stop in San Juan (Puerto Rico). The most extensive inter-range services are known as "round-the-world" routes as major marine ranges of the world are services along a continuous loop. Another recent trend has been the integration and specialization of several routes with feeder ships converging at major marine intermediate hubs. This is notably the case for Europe (Mediterranean, North Sea, and the Baltic) in light of the negative impacts of **deviations** from main marine shipping routes in terms of service length and frequency of port calls.

(1,491 words)

➢ *New Words*

intensification [ɪnˌtensɪfɪ'keɪʃn]	*n.* action that makes something stronger or more extreme 强化;加剧;激烈化
	WHO is providing technical support to the Ministry of Health, including the intensification of surveillance and case management.
	世卫组织正在向卫生部提供技术支持,包括加强监测和病例管理。
exponentially [ˌekspə'nenʃəlɪ]	*ad.* growing or increasing very rapidly 以指数方式;呈指数地
	The government says that the social security budget will rise exponentially.
	政府声称社会保障预算将会呈指数上升。
reciprocity [ˌresɪ'prɒsətɪ]	*n.* mutual exchange of commercial or other privileges 互惠
	On some occasions, some countries allow entry of foreigners without visas through agreement on the basis of reciprocity.
	某些情况下,有些国家会在互惠的基础上达成协议,互免签证。
attribute ['ætrɪˌbjuːt]	*n.* an abstraction belonging to or characteristic of an entity 属性;特性
	Another key attribute of this smart phone is that it can record up to 10 minutes of video.
	这款智能手机的另一个主要特点就是可以录制长达 10 分钟的视频。
physiography [ˌfɪzɪ'ɒgrəfɪ]	*n.* the study of physical features of the earth's surface 地文学,地相学
	The petroleum exploration geologist working with subsurface channel sandstone must have a background knowledge of physiography.
	石油地质工作者经常要和地下的水道砂岩打交道,必须具备地文学方面的背景知识。
itinerary [aɪ'tɪnərərɪ]	*n.* an established line of travel or access 旅程;路线
	Shanghai is the home port of three cruises from both Costa and Royal Caribbean, and is a fixture on the itinerary of many round-the-world cruises.
	目前,上海港不仅是歌诗达和皇家加勒比这两家公司

三艘游轮的母港,也是很多环球游轮的固定停泊港口。

discontinuity [ˌdɪsˌkɒntɪˈnjuːətɪ]　　*n.* lack of connection or continuity 不连续;中断;间断性

Changes in government led to discontinuities in policy.

政府的更迭导致政策缺乏连续性。

dredge [dredʒ]　　*v.* to remove mud and unwanted material (usually from a bottom of a body of water) with a special machine in order to make it deeper or to look for something 疏浚(河道等);清淤;挖掘

When we dredge the bottom and bring up the remains of animal and vegetable life we find that they give evidence of not having been disturbed in the least, for hundreds and thousands of years.

从海底挖掘出来的动植物遗迹向我们表明:它们埋在海底数百年数千年之久,从来没有受到丝毫干扰。

massification [mæsɪfɪˈkeɪʃən]　　*n.* in a popular style 大众化

Higher education massification refers to the evolution process of cultivating the minority of elite to the majority of the society and then to all the people, and the corresponding qualitative changes.

高等教育大众化是指高等教育从培养少数精英到面向社会大多数人直至全体的发展历程以及由此而产生的质变。

fluvial [ˈfluːvɪəl]　　*a.* of or relating to or happening in a river 河流的;生在河中的

The research in fluvial sedimentology is significant not only for theory but also for practice of production and flood controlling.

河流沉积学的研究不但具有重要的理论意义,而且对生产实践和洪灾防治具有重要的指导作用。

interface [ˈɪntəfeɪs]　　*n.* the overlap where two theories or phenomena affect each other or have links with each other 接口;界面

The adapter supports outbound mode only for this interface.

对于此接口,适配器仅支持输出模式。

infrastructure [ˈɪnfrəstrʌktʃə(r)]　　*n.* the basic structure or features of a system or organization 基础设施;公共建设;下部构造

Although South Africa has many of the attributes of the first world—some good infrastructure, millions of rich people—it is still not part of that world.

虽然南非有着很多第一世界的特征——有些不错的基础设施，还有不计其数的富人，但依然不属于第一世界。

hydrographic [ˌhaɪdrəʊˈgræfɪk]

a. of or relating to the science of hydrography 与水道测量有关的；水道学的

Back in 1795, the increased need for reliable charts during the French Revolution led to the development of the Hydrographic Office of the Admiralty.

1795 年法国大革命期间，对海图的准确性要求越来越高，英国海军水道测量局随之发展起来。

artery [ˈɑːtərɪ]

n. a major thoroughfare that bears important traffic 干道；主流

As well as providing a short cut for battleships, the canal became a vital artery of world trade.

该运河不仅为战舰提供了一条近路，还成为世界贸易的一条主干道。

enclave [ˈenkleɪv]

n. an area within a country or a city where people live who have a different nationality or culture from the people living in the surrounding country or city 被包围的领土；飞地（指在本国境内的隶属另一国的一块领土）

The question now is whether this right applies only in a federal enclave such as Washington, DC, or nationwide.

现在的问题是，这一权利是否仅适用于华盛顿特区这样的联邦特区，还是全国范围都适用。

nautical [ˈnɔːtɪkl]

a. relating to or involving ships or shipping or navigation or seamen 航海的，船上的；船员的

Until 500 years ago, our ancestors still led advancements in fields of nautical navigation, ceramics and silk.

直至 500 多年前，我们的祖先在航海、陶瓷、丝织等诸多领域还处于世界领先地位。

configuration [kənˌfɪgəˈreɪʃn]

n. an arrangement of parts or elements 配置；结构；外形

Prices range from $119 to $199, depending on the particular configuration.

配置不同，价格不同，从 119 美元到 199 美元不等。

backhaul [ˈbækhɔːl]

n. haul cargo back from point B to the originating point A 载货回程；回程运费；回运

There are three major constraints on streaming video to a mobile device over the air：wireless spectrum, backhaul, and the device itself.

无线传输流媒体视频到移动设备,存在三大局限:无线带宽、信息回传以及设备本身的限制。

deviation [ˌdiːvɪˈeɪʃn]　　　　　*n.* a variation that deviates from the standard or norm 偏离;误差;背离

A minimal error or deviation may result in wide divergence.

失之毫厘,谬以千里。

➢ *Phrases and Expressions*

be subject to 受支配,从属于;常遭受……;有……倾向的

in conjunction with 连同,共同;与……协力

account for 对……负有责任;对……做出解释;说明……的原因;导致

in terms of 依据;按照;在……方面;以……措词

be composed of 由……组成

rely on 依靠,依赖

be relevant to 和……相关

be opposed to 反对……;与……相对

prior to 在……之前;居先

in light of 根据;鉴于;从……的观点

➢ *Terminology*

marine route 海运航线

marine current 洋流,即海流,也称洋面流,是指海水沿着一定方向有规律的具有相对稳定速度的水平流动,是从一个海区水平或垂直地向另一个海区大规模的非周期性的运动,是海水的主要运动形式

container ship 集装箱船

bulk carrier 散货船

fossil fuel 矿物燃料,也称化石燃料,是一种烃或烃的衍生物的混合物,包括煤炭、石油和天然气等天然资源,是不可再生资源

fluvial-marine ship 江海两用船

break-bulk ship 杂货船

port call 沿途到港停靠

load factor 载运率

feeder ship 支线船

➤ *Proper Names*

BCE 公元前

Sumatra 苏门答腊岛

Mediterranean 地中海

Caribbean 加勒比海

Suez Canal 苏伊士运河

Amoco Cadiz 指代"卡迪兹号"油轮沉没事件,1978 年 3 月 16 日,利比里亚籍超大型油轮"卡迪兹号"在英吉利海峡靠法国一侧航行时遇强风偏航导致触礁沉没,泄漏出 23 万吨石油并且污染了附近整个海面和法国海岸

Exxon Valdez 指代"埃克森-瓦尔迪兹号"漏油事件,1989 年 3 月 24 日,美国埃克森公司巨型油轮"瓦尔迪兹号"在阿拉斯加州美、加交界的威廉王子湾附近触礁,原油泄出达 800 多万加仑,在海面上形成一条宽约 1 千米、长达 800 千米的漂油带。事故发生地点是一个原来风景如画的地方,盛产鱼类,海豚海豹成群。事故发生后,礁石上沾满一层黑乎乎的油污,不少鱼类死亡,附近海域的水产业受到很大损失,生态环境遭受巨大的破坏

Antwerp 安特卫普,比利时最大港口和重要工业城市

Sea-Land 美国海陆联运公司,前身为美国泛大西洋轮船公司,1960 年更名为海陆联运公司,1999 年被马士基集团(旗下的马士基航运是全球最大的集装箱运输公司)兼并。1956 年 4 月,美国泛大西洋轮船公司在一艘 T-2 型油船甲板上设置了一个可装载 58 只 35 英尺集装箱的平台,取名"马科斯顿号",航行于纽约至休斯敦航线上。1957 年 10 月,该公司又将一艘 C-2 型货船改装成吊装式全集装箱船,取名"盖脱威城号",载重量 9 000 t,可装载 226 个 35ft 集装箱,这是世界上第一艘开展海上运输的集装箱船。1960 年 4 月,为了突出集装箱的联运特点,泛大西洋船公司改名为海陆联运公司。1961 年,该公司陆续开辟了纽约—洛杉矶—旧金山航线和阿拉斯加航线,奠定了美国集装箱运输的基础

Panama Canal 巴拿马运河

Puerto Rico 波多黎各

Baltic 波罗的海

➤ *Translation*

1. Thus, access to trade commodities remains historically and... (Para. 2)
   因此,不管是过去还是现在,人们建立海上运输网,最主要的动机还是进行商品贸易。

2. With the development of the steam engine in the mid-19th century... (Para. 3)
   19 世纪中叶,随着蒸汽机的进一步发展,船舶行驶不再受制于风力、风向,海上贸易网得以大大扩展。

3. These trade relations are also influenced by the existing marine... (Para. 3)
   现有的海上运力状况以及海上运输服务体系变化,都会对这些贸易关系造成影响。

4. Marine transportation, like land and air modes, operates on its... (Para. 4)

与陆地运输和航空运输一样,海上运输也有自己的专属领地,就自然属性而言,有其地理优势,从运营控制上讲,有其战略意义,就资源利用来说,有其商业价值。

【英汉翻译中常常在形容词前后增译名词使其意义更加完整,如此例中将"geographical"译为"地理"后增译"优势",将"strategic"译为"战略"后增译"意义",将"commercial"译为"商业"后增译"价值";另外,考虑到原文是平行句式,译文中也要注意行文排比,尽量再现源语的句式特点。】

5. The massification of transport into regular flows over long distances... (Para. 6)

运输业务覆盖越来越广,出现长途定期专线,也有可能带来一些不良后果,比如一旦有些油轮发生意外,就可能导致重大的生态灾难(例如"卡迪兹号"油轮沉没事件和"埃克森瓦尔迪兹号"漏油事件)。

(1978年3月16日,利比里亚籍超大型油轮"卡迪兹号"在英吉利海峡靠法国一侧航行时遇强风偏航导致触礁沉没,泄漏出23万吨石油并且污染了附近整个海面和法国海岸。1989年3月24日,美国埃克森公司巨型油轮"瓦尔迪兹号"在阿拉斯加州美、加交界的威廉王子湾附近触礁,原油泄出达800多万加仑,在海面上形成一条宽约1千米、长达800千米的漂油带,礁石上沾满一层油污,不少鱼类死亡,生态环境遭受巨大破坏。)

6. Ports are less relevant to fluvial transportation but fluvial hub centers... (Para. 7)

港口情况对河运影响不大,但有些河运枢纽与海运和陆运,尤其是集装箱运输的关系越来越紧密。

7. Despite regular services on selected fluvial arteries, such a... (Para. 8)

尽管某些河运干道(例如长江)也有定期班轮运行,但河运客轮的开发前景,目前仍仅限于旅游观光(内河游轮)。

8. Significant investments have been made in expanding transshipment... (Para. 8)

为了提高港口转运能力,已经投入了大量资金,不过,由于港口占地面积很大,这种建设项目也是相当烧钱的。

【科技英语重在事物描述,强调行文客观,很少涉及动作实施者,所以多用被动语态。相比之下,被动句在汉语中的使用频率不高,所以在英汉翻译中,常将被动语态改译为主动语态,或者增译虚指主语如"人们"、"大家",或者译成无主句,此例中采用第二种技巧;另外,增译"这种建设项目"来指代"which"引导的定语从句所修饰的先行词,使上下文更加通顺。】

9. However, being landlocked does not necessarily imply exclusion... (Para. 9)

但是,没有海港并不一定意味着无法进行国际贸易,只是运输成本会大大提高,这可能妨碍经济发展。

10. Prior to containerization, loading or unloading a ship was a very... (Para. 10)

没有实现集装箱化以前,装船、卸船既耗时又耗资,货船在海上航行的时间都没有等在码头上装货、卸货的时间长。

11. Another recent trend has been the integration and specialization... (Para. 12)

近年来,还有另外一个趋势,就是把有些航线整合起来,开通支线船在几个海运枢纽集中运营,打造货运专线。

【英语是静态语势,使用名词较多,汉语是动态语势,使用动词较多,英汉翻译中,常将名词转译为动词,如"integration"和"specialization"都转译成了动词;另外,英语重形合,多长句、复杂句,而汉语重意合,多短句、分句、流水句,因此在翻译时常将英语的长句拆分成几个短句。】

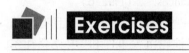

Exercises

1. Fill in the blanks with the proper given words, and then translate the sentences into Chinese.

dominant	represent	perceive	strategic
access	be subject to	account for	in terms of
be relevant to	inflexible	integration	privilege
significant	sufficient	prior to	in light of

1) You should bear in mind that a good many things in the world cannot be considered and valued _____ money although we have to admit that it is very important.

2) All in all, at least at this stage, it's unlikely that online learning will become the _____ form of education.

3) The developments in this town _____ in microcosm what is happening in the country as a whole.

4) The United States and Canada may enter into an agreement that would allow easier _____ to jobs across the border.

5) How you face and _____ these challenges, though, will determine your stress and anxiety levels throughout this process.

6) Its geographical position has given it a _____ importance, especially in the energy-pipeline business.

7) Meanwhile, the deterioration of trade conditions also _____ the continual upgrading of the trade frictions.

8) Just like America during its take-off in the late 19th century, emerging economies tend to _____ economic ups and downs.

9) There is _____ evidence to show that plastic bags have caused white pollution to the environment.

10) Under the law, prosecutors are obligated to reveal any and all information that could _____ the guilt of the defendant, particularly information that would be exculpatory.

11) Every country has the right to choose, _____ its national conditions, its social system and

road to development independently.

12）A very determined effort by society will ensure that the disabled achieve real acceptance and _____.

13）His opponents viewed him as stubborn and _____ while his colleagues thought of him as determined and responsible.

14）The mass media have been vested with _____ power as social and political agents in modern developed societies.

15）The Russian Federation has issued a decree abolishing special _____ for government officials.

16）According to most of the historians, _____ the nineteenth century, there were almost no channels of social mobility.

2. Translate the following passage into Chinese.

Marine transportation concerns the movement of passengers and freight over water masses, from oceans to rivers. From its modest origins as Egyptian coastal and river sailships around 3,200 BCE, marine transportation has always been the dominant support of global trade. By 1,200 BCE Egyptian ships traded as far as Sumatra, representing one of the longest marine routes of that time. By the 10th century, Chinese merchants frequented the South China Sea and the Indian Ocean, establishing regional trade networks. In the early 15th century, Admiral Zheng He led a large fleet of more than 300 vessels manned by a crew of 28,000 to conduct seven major expeditions, one of which reached the East African coast. However, China's attempt at asserting a regional marine dominance was short-lived and such expeditions were not permitted to continue mostly because China perceived itself as a continental power with marine trade of limited interest.

3. Translate the following passage into English.

15 世纪初，一支船队从南京启航，揭开了中国海上航行的序幕，这支船队是由中国历史上最伟大的探险家，也是举世闻名的航海家郑和率领的。在 28 年的时间里，他先后 7 次率队进行海上远征，探访了 40 多个国家，打开了中国在非洲、印度及东南亚的贸易航线。郑和于公元 1433 年去世，他死后 3 年，新帝登基，之后突然下诏禁止再建造远洋船只，中国的海上扩张时代就此戛然而止。究竟为何会发生这样的变故，历史上一直众说纷纭。不管什么原因，这一新政使中国将大海拱手让给了欧洲后起之国，古老华夏自此再也未能称霸海上。

Unit 2
Interoceanic Passages and Seaports

Interoceanic Passages and Land Corridors

1. Marine transportation is the main support of international freight distribution and e-volves over a global marine space. This space has its own **constraints** such as the **profile** of continental masses and the **detours** and passages it creates. Therefore, **interoceanic** canals are built to improve the connectivity of shipping networks when possible. The most important strategic marine passages are known as **chokepoints** (or bottlenecks) due to:

2. Capacity constraints. Chokepoints tend to be shallow and narrow, **impairing** navigation and imposing capacity limits on ships. For interoceanic canals such as Panama and Suez, the capacity must effectively be managed with appointment and pricing systems.

3. Potential for **disruptions** or **closure**. Disruption of trade flows through any of these routes could have a significant impact on the world economy. Many chokepoints are next to politically unstable countries, increasing the risk of **compromising** their access and use, such as with **piracy**. Closures are a rare instance that only took place in situations of war as one **proponent** prevented another to access and use the chokepoint (e. g. Gibraltar and Suez during World War II). Closure of a marine chokepoint in the current global economy, even if temporary, would have important economic consequences with the disruption of trade flows and even the interruption of some supply chains (e. g. oil). These potential risks and impacts are commonly used to justify military naval **assets** to protect sea lanes even if such benefits are difficult to demonstrate.

4. Changes in the technical and operational characteristics of interoceanic canals and

passages can have substantial impacts on global trade patterns. The Panama Canal, the Suez Canal, the Strait of Malacca and the Strait of Hormuz account for the world's four most important interoceanic passages. This is in part because of the chokepoints they impose on global freight circulation and in part because of the economic activities and resources they grant more efficient access to. Their continuous availability for global marine trade is challenging because the global trade system is highly **reliant** on their use. Yet, they have shaped global trade with the ongoing setting of rings of circulation comprised of marine and land corridors, notably in the northern hemisphere.

5. The Suez Canal is an artificial waterway of about 190 km in length running across the **Isthmus** of Suez in northeastern Egypt which connects the Mediterranean Sea with the Gulf of Suez, an arm of the Red Sea. The Suez Canal has no locks because the Mediterranean Sea and the Gulf of Suez have roughly the same water level, which makes Suez the world's longest canal without locks. It acts as a shortcut for ships between both European and American ports and ports located in southern Asia, eastern Africa, and Oceania. Because of obvious geographical considerations, the marine route from Europe to the Indian and Pacific oceans must round the Cape of Good Hope at the southernmost point of the African continent. There are a number of alternatives to the Suez Canal but they involve either very long detours or have limited capacity.

6. The Panama Canal joins the Atlantic and Pacific oceans for 82 kilometers across the Isthmus of Panama, running from Cristobal on Limon Bay, an arm of the Caribbean Sea, to Balboa, on the Gulf of Panama. Since its expansion in 2016, it involves two systems of locks that can both be used at the same time. The old locks, completed in 1914, can handle ships with a draft of 12. 2 meters (40 feet), a width of 32 meters (106 feet) and a length of 294 meters (965 feet). The expanded locks, completed in 2016, can handle ships with a draft of 15. 2 meters (50 feet), a width of 49 meters (160 feet) and a length of 366 meters (1200 feet).

7. The Strait of Malacca is one of the most important strategic passages of the world because it supports the bulk of the marine trade between Europe and Pacific Asia, which accounts for 50,000 ships per year. About 30% of the world's trade and 80% of Japan's, South Korea's and Taiwan's imports of petroleum **transits** through the strait, which involved approximately 15. 2 Mbd in 2013. It is the main passage between the Pacific and the Indian oceans with the strait of Sunda (Indonesia) being the closest

existing alternative. It measures about 800 km in length, has a width between 50 and 320 km (2.5 km at its narrowest point) and a minimum channel depth of 23 meters (about 70 feet). It represents the longest strait in the world used for international navigation and can be transited in about 20 hours.

8.   The Strait of Hormuz forms a strategic link between the oil fields of the Persian Gulf, which is a marine dead-end, the Gulf of Oman and the Indian Ocean. It has a width between 48 and 80 km, but navigation is limited to two 3 km wide channels, each exclusively used for **inbound** or **outbound** traffic. Circulation in and out of the Persian Gulf is thus highly constrained, namely because the sizable amount of tanker and containership traffic makes navigation difficult along the narrow channels. In addition, islands that ensure the control of the strait are **contested** by Iran and the United Arab Emirates. The strait is deep enough to accommodate all the existing tanker classes.

9.   Besides the four passages mentioned above, there are other important passages:

10.   The Strait of Bab el-Mandab? is controlling access to the Suez Canal, a strategic link between the Indian Ocean and the Red Sea. Its width is only 18 miles wide at its narrowest point. The majority of southbound traffic passing through Suez Canal must also pass through Bab el-Mandeb. As of 2016, about 4.8 billion barrels of oil passed through this passage. Closing of this strait would have serious consequences, forcing a detour around the Cape of Good Hope and in the process demanding additional tanker space. Like the Strait of Malacca, Bal el-Mandab is a crucial link in the Europe—Asia trade route.

11.   The Oresund Strait is a passage of 115 km between Denmark and Sweden connecting the North Sea and the Baltic, which is a marine dead-end. It enables Russia, the Baltic States, Poland, and Germany to access international marine shipping. The majority of Russian container trade transit through the strait.

12.   The Strait of Gibraltar is a **peninsula** between the Atlantic and the Mediterranean oceans and represents an obligatory passage point between these two oceans. The strait is about 64 km long and varies in width from 13 to 39 km. The Strait has become an important transshipment **nexus** along major Europe / Mediterranean / Asia shipping routes. More recently, the growth of the trade using the Suez Canal to reach the North American East Coast has provided additional transshipment opportunities around the strait.

13. The Strait of Bosporus has a length of 30 km by of width of only 1 km at its narrowest point linking the Black Sea to the Mediterranean Ocean. With the passage of the Dardanelles, Bosporus forms the only link between the Black Sea and the Mediterranean Ocean. In the current context, Bosporus represents a passage of growing strategic importance, notably after the fall of the Soviet Union.

14. The Strait of Magellan? was crossed in 1520 by the Portuguese explorer Ferdinand Magellan and separates South America to Tierra del Fuego with a length of 530 km long and a width of 4 to 24 km. It provides a better-protected passage than the Drake Passage, which is the body of water between South America and Antarctica. It was initially held secret for more than one century to assure the **supremacy** of Portugal and Spain for the Asian trade of spices and silk. With the construction of the Panama Canal in 1916 and later on the setting up of the North American transcontinental bridge in the 1980s, this passage has lost most of its strategic importance.

15. The Cape Good Hope represents the extreme tip of Africa separating the Atlantic and Indian oceans and was transited by the Portuguese at the end of the fifteenth century. It took its name because of the fact that it offered a marine passage towards India and Asia, thus the hope of a fortune for the one who passed it. Since the widening of the Suez Canal in the 1970s, the Cape of Good Hope has lost some of its strategic importance but still remains an important passage.

(1,406 words)

➤ *New Words*

constraint [kənˈstreɪnt]	*n.*	a thing that limits or restricts something, or your freedom to do something 限制；约束
		Their decision to abandon the trip was made because of financial constraints.
		因经济条件所限，他们决定放弃这次旅行。
profile [ˈprəʊfaɪl]	*n.*	the edge or outline of something 侧面；轮廓；外形
		We could see the profile of a distant hill if it is very clear.
		如果天气晴朗，我们可以看到远山的轮廓。
detour [ˈdiːtʊə(r)]	*n.*	a roundabout road (especially one that is used temporarily while a main route is blocked) 迂回路；临时绕行道路
		He did not take the direct route to his office, but made a detour around the outskirts of the city.
		他没有直接上班，而是在市郊绕了一段路。
interoceanic [ˌɪntərəʊʃɪˈænɪk]	*a.*	connecting two oceans 大洋间的；连接两大洋的
		Brazil has shied away from financing South American development banks, but has enthusiastically worked to finance the construction of the Interoceanic Highway, a freeway that links Brazil to Pacific coast ports in Peru.
		巴西不愿为南美开发银行提供资金，但热衷于投资跨洋高速公路的建设，是因为这条路建好之后，可以连通巴西和秘鲁的太平洋沿岸港口。
chokepoint [ˈtʃəʊkˌpɔɪnt]	*n.*	a point of congestion or blockage 阻塞点；交通枢纽点
		Iran granted them units basing rights in the strategically vital Bab al-Mandab strait that controls the chokepoint connecting the Indian Ocean with the Red Sea.
		伊朗同意他们在曼德海峡驻军，该海峡是印度洋和红海之间的咽喉要道，战略位置极其重要。
impair [ɪmˈpeə(r)]	*v.*	to make worse or less effective 损害；削弱；减少
		While overindulging can make the brain sluggish, too few calories can also impair brain function.
		虽然暴饮暴食会让大脑反应迟钝，但是摄入卡路里太少也会损害大脑功能。
disruption [dɪsˈrʌpʃn]	*n.*	an act of delaying or interrupting the continuity 扰乱，打乱，中断
		In 2009, in response to soaring prices, food riots and disruption to world trade, rich governments also began to wor-

ry about agriculture, and promised to do something about its problems.

2009 年,物价飞涨,粮食供应不稳,世界贸易形势也不乐观,因此,发达国家政府也开始担心农业发展状况,并承诺将采取措施解决问题。

closure [ˈkləʊʒə(r)]　　*n.* the permanent ending of something 关闭;终止,结束

Restaurant and club owners, and employees have been protesting the closure of their workplaces outside of government offices, arguing that the measures taken to prevent the spread of the virus are depriving them of their livelihood.

餐馆和酒吧老板及其员工一直守在政府办公楼外,抗议政府关闭他们的工作场所。他们认为,为防止病毒传播而采取的这些措施让他们无法维持生计。

compromise [ˈkɒmprəmaɪz]　　*v.* to reach an agreement by giving up something 妥协,折中;使陷入危险,名誉受损

Every time someone tells you to "be realistic" they are asking you to compromise your ideals.

有人告诉你"现实一点"的时候,其实是在让你妥协,放弃自己的理想。

piracy [ˈpaɪrəsɪ]　　*n.* robbery at sea carried out by pirates 海盗行为;剽窃;著作权侵害

As stated in its defense white paper this year, the country pays increasing attention to non-traditional security threats such as terrorism and piracy.

该国在今年的国防白皮书中声称越来越关注恐怖主义和海盗等非传统安全威胁。

proponent [prəˈpəʊnənt]　　*n.* a person who pleads for a cause or propounds an idea 支持者;建议者

He was identified as a leading proponent of the values of progressive education.

他被认为是进步教育价值观的倡导者。

asset [ˈæset]　　*n.* a useful or valuable quality 资产;优点;有用的东西;有利条件

The greatest asset is our creativity in the field of technology.

最大的资本就是我们在技术领域的创造力。

reliant [rɪˈlaɪənt]　　*a.* relying on another for support 依赖的;可靠的;信赖的

Asia must burn more fuel to generate economic growth, primarily because it is more reliant on energy-intensive industries than other parts of the world.

比起世界其他地区,亚洲对于高能耗产业更加依赖,所以需要消耗更多的能源以便推动经济增长。

isthmus [ˈɪsməs]   *n.* a narrow piece of land connecting two large areas of land 地峡;管峡

Auckland, located on an isthmus of northwest North Island, is the largest city of New Zealand, a major port and an industrial center.

奥克兰位于北岛西北部地峡,是新西兰最大的城市,也是主要海港及工业中心。

transit [ˈtrænzɪt]   *v.* to make a passage or journey from one place to another 运输;经过

Russia and Italy signed an agreement that will allow military materials to transit Russian territory.

俄罗斯和意大利签署协议,允许军用物资过境俄罗斯领土。

inbound [ˈɪnbaʊnd]   *a.* directed or moving inward or toward a center 归航的;到达的

Japanese inbound flights from Shanghai were mostly normal over the weekend, despite the disaster.

尽管发生了灾难,但上周末从上海返回的日本航班基本正常。

outbound [ˈaʊtbaʊnd]   *a.* going out or leaving 向外的;出港的;离开某地的

The company is on good terms with many outbound investment and financing institutes and securities dealers, which can offer assistance to the investment companies in time.

公司与国外多家投融资机构以及证券商有着良好关系,可为投资企业提供及时帮助。

contest [kənˈtest]   *v.* to make the subject of dispute, contention, or litigation 就……提出异议,反驳;争取赢得(比赛、选举等);参加(竞争或选举),竞争;争辩,争论

Your former employer has to reply within 14 days in order to contest the case.

你的前雇主必须在14天内做出答复以对本案提出抗辩。

peninsula [pəˈnɪnsjələ]   *n.* a large mass of land projecting into a body of water 半岛

Dalian is in the south of the Liaodong Peninsula.

大连位于辽东半岛南部。

nexus [ˈneksəs]   *n.* the means of connection between things linked in series 关系;连结

Together they form a complex nexus of mutually reinforcing, intertwined patterns.

他们以相互促进、相互交叉的方式构成了一种复杂的关系。

supremacy [suːˈpreməsɪ]   *n.*   power to dominate or defeat 霸权；至高无上；主权；最高地位

The United States has managed to maintain its supremacy in cotton by investing heavily in technology.

美国在技术方面投入大量资金，以此维持其在棉花市场上的霸主地位。

➤ *Phrases and Expressions*

due to 由于；应归因于
have an impact on. 对······有影响
impose on 利用；欺骗；施加影响于
in part 部分地；在某种程度上
be comprised of 由······组成
in addition 另外，此外
be limited to 被限制在

➤ *Terminology*

lock 船闸，厢形构筑物，由上、下游引航道与上、下游闸首连闸室组成。闸室是停泊船舶（或船队）的厢形室，借助室内灌水或泄水来调整闸室中的水位，使船舶在上、下游水位之间作垂直的升降，从而通过集中的航道水位落差。当船舶由下游向上游行驶时，室内水位降至与下游水位齐平，然后打开下游闸首的闸门，船进闸室，关闸门，灌水，待水位升高到与上游水位齐平后，开上游闸首闸门，船即可出闸通过上游引航道驶向上游。当船由上游向下游行驶时，过闸操作程序则与此相反。

MBD 国际标准原油计量单位，Thousands of Barrels Daily 指千桶每日；另一种解释，Million Barrels per Day，即百万桶每日。很多国家石油协会的统计报表用的是"千桶每日"，针对输油管道，MBD 只要不加注明，一般指千桶每日，反倒是表示百万桶每日时要写清楚 million barrels per day（MBD）。

➤ *Proper Names*

Gibraltar 直布罗陀（海峡）
Strait of Malacca 马六甲海峡
Strait of Hormuz 霍尔木兹海峡
Cape of Good Hope 好望角

Cristobal 克里斯托瓦尔(巴拿马最大港口)

Limon Bay 利蒙湾(加勒比海的天然港湾,巴拿马运河北端)

Balboa 巴尔博亚(巴拿马运河区内港市)

Sunda 巽他(海峡)

Gulf of Oman 阿曼湾

United Arab Emirates 阿拉伯联合酋长国

Strait of Bab el-Mandab 曼德海峡

Oresund Strait 厄勒海峡

Strait of Bosporus 博斯普鲁斯海峡

Dardanelles 达达尼尔海峡

Strait of Magellan 麦哲伦海峡

Tierra del Fuego 火地岛

Drake Passage 德雷克海峡

➤ *Translation*

1. This space has its own constraints such as the profile of... (Para. 1)

(当然),这个空间也有其自身限制,例如大陆板块形状各异、犬牙交错,形成了各种弯弯曲曲的通道。

【原文中"such as"后面是三个并列的名词结构,后两个名词结构还带有一个定语从句"it creates",如果按照原文的词性和结构翻译,既不符合汉语措辞习惯又不符合其行文习惯,翻译时应将其拆分为多个短句,因此首先根据上下文增译"各异、犬牙交错"组成第一个分句,其次将原文中的定语从句"it creates"及其所修饰的先行词"detours and passages"合译成为第二个分句,使译文表达更加清楚、流畅。】

2. Chokepoints tend to be shallow and narrow, impairing navigation... (Para. 2)

这种咽喉要道,往往水浅路窄,航行极其不便,而且对通行船只的货运量也有限制。

3. Many chokepoints are next to politically unstable countries... (Para. 3)

很多咽喉要道都毗邻政局不稳的国家,这会大大增加通行风险,比如可能要和海盗谈判,争取通行权利。

4. The Suez Canal has no locks because the Mediterranean Sea... (Para. 5)

地中海和苏伊士湾的水位大致相同,所以苏伊士运河没有船闸,这使苏伊士运河成了世界上最长的无船闸运河。

5. The Strait of Malacca is one of the most important strategic... (Para. 7)

欧洲与亚太地区之间的海上贸易大多取道马六甲海峡,每年通行船只多达 50 000 艘,因此,马六甲海峡也是世界上最重要的战略要道之一。

【英语句子信息呈现顺序多是先中心后外围,而汉语则是先外围后中心,一般按照由先到后、由因到果、由事实到结论的顺序铺排信息,所以英汉翻译中常采取逆序翻译,以符合汉语的思维方式和表达习惯。】

6. Circulation in and out of the Persian Gulf is thus highly. . . （Para. 8）

（因此，）进出波斯湾非常受限，因为航道本来就很狭窄，又有大量油轮和集装箱船，使通行更加困难。

7. Closing of this strait would have serious consequences, forcing. . . （Para. 10）

该海峡一旦断航，会导致严重后果，来往船只将不得不从好望角绕行，如此一来就需要加大运量。

8. More recently, the growth of the trade using the Suez Canal. . . （Para. 12）

最近，取道苏伊士运河抵达北美东海岸的海上贸易有增长趋势，如此一来，船运公司比以前更有可能选择在直布罗陀海峡周边码头进行转运。

9. In the current context, Bosporus represents a passage of growing. . . （Para. 13）

在当前局势下，特别是苏联解体之后，博斯普鲁斯海峡（又称伊斯坦布尔海峡）的战略地位越来越重要了。

10. It was initially held secret for more than one century to assure. . . （Para. 14）

在一个多世纪的时间里，为了保证葡萄牙和西班牙能够优先与亚洲进行贸易活动，获得所需的香料和丝绸，这条通道一直没有被公之于世。

【首先，原文中"supremacy"是一个翻译难点，翻译时先将其从主句中拆分开来，同时将它和"trade"都转译成为动词，扩充成为一个分句；其次，原文是英语典型的多支共干结构，信息从主干到分支逐级展开，而汉语则是板块式结构，按语义或事理逻辑铺排，没有一定的框架限制，多流水句和松散句，所以在英汉翻译中经常使用拆分技巧，将原文拆分为多个短句；另外，通过正反对译，将原文中的肯定表达"held secret"译为否定表达"没有被公之于世"，既符合汉语表达习惯，又突出原文想要表达的神秘色彩。】

11. It took its name because of the fact that it offered a marine. . . （Para. 15）

好望角之所以得名，是因为它是通向印度和亚洲的海上通道，传说经过这里的人会有望交到好运。

 **Exercises**

1. Fill in the blanks with the proper given words, and then translate the sentences into Chinese.

distribution	impair	potential	disruption
have an impact on	compromise	justify	impose on
grant	challenging	be comprised of	exclusively
contest	accommodate	crucial	initially

1) There is a(an) _____ educational benefit in allowing pictures to tell the story, rather than the spoken word.

2) Some of the victims are complaining loudly about the uneven _____ of emergency aid.

3) It is hard to say if the movement will ultimately _____ the issues of racial discrimination and income inequality.

4) Although it is _____ to work in Antarctica, scientists' passion to learn about this mysterious land will never end.

5) Inadequate sleep is known to _____ the ability to think, handle stress, maintain a healthy immune system, and keep emotions in check.

6) _____ to trade finance means that developing world suppliers are having trouble getting goods to market.

7) They called for a _____ on all sides to break the deadlock in the world trade talks and move forward for mutual benefit.

8) The death toll in the earthquake was _____ reported at around 250, but was later revised to 300.

9) It is universally acknowledged that no excuse can _____ one country's intervention in the internal affairs of another country.

10) The suite used to _____ three separate bedrooms, a layout the current owner has changed for more space.

11) Strengthening cooperation can greatly lighten the restriction that the land dimensions _____ the scaled economy.

12) It _____ me the serenity to accept the things I cannot change, the courage to change the things I can change and wisdom to know the difference.

13) The proposals to the legislature include the creation of two special courts to deal _____ with violent crimes.

14) The oil company, which has already paid penalties, said it will _____ the charge in the next month.

15）The court overturned that decision on the grounds that the Prosecution had withheld _____ evidence.

16）The library can _____ around 100 children, some of whom come to do their homework after school or to read in the evenings.

2. Translate the following passage into Chinese.

Changes in the technical and operational characteristics of interoceanic canals and passages can have substantial impacts on global trade patterns. The Panama Canal, the Suez Canal, the Strait of Malacca and the Strait of Hormuz account for the world's four most important interoceanic passages. This is in part because of the chokepoints they impose on global freight circulation and in part because of the economic activities and resources they grant more efficient access to. Their continuous availability for global marine trade is challenging because the global trade system is highly reliant on their use. Yet, they have shaped global trade with the ongoing setting of rings of circulation comprised of marine and land corridors, notably in the northern hemisphere.

3. Translate the following passage into English.

埃及宣布全国放假一天,庆祝新苏伊士运河开通。开通仪式上,一支舰队沿新河道通行,其中还包括 1869 年苏伊士运河开通时首支通过该运河的游艇。苏伊士运河于 1869 年完工,被称为阿拉伯的"水上地铁"。此次扩建工程长度约为 45 英里,是运河总长的三分之一。扩建之前,运河无法双向通行,船队只能单向鱼贯而过。现在,双向船队可以同时通过运河,每趟航行可以节省 8 个小时的时间,成本大大降低。苏伊士运河管理部门表示,预计到 2023 年,经航船只数量将会翻倍,运河年收入可增至 130 亿美元。

Unit 3
Marine Shipping

Marine Shipping Services and Networks

1. The marine shipping industry is part of a life cycle that includes building, registration, operations and the final **scrapping** of the ship. Marine shipping is dominated by bulk cargo, which roughly accounted for 69.6% of all the ton-miles shipped in 2005. However, the share of break-bulk cargo is increasing steadily, a trend mainly attributed to containerization. Marine shipping has traditionally faced two **drawbacks** in relation to other modes. First, it is slow, with speeds at sea averaging 15 knots for bulk ships (26 km/h, 1 knot = 1 marine mile = 1,853 meters), although container ships are designed to sail at speeds above 20 knots (37 km/h). Secondly, delays are encountered in ports where loading and unloading take place. The latter may involve several days of handling when break-bulk cargo was concerned. These drawbacks are particularly constraining where goods have to be moved over short distances or where shippers require rapid deliveries.

2. Marine shipping has seen several major technical innovations aiming at improving the performance of ships or their access to port facilities, notably in the 20th century. They include:

3. Size. The last century has seen growth in the number of ships as well as their average size. Each time the size of a ship is doubled, its capacity is cubed (tripled). Although the minimum size for cost-effective bulk handling is estimated to be around 1,000 deadweight tons, economies of scale have pushed for larger ship sizes to service transportation demand. For ship owners, the rationale for larger ships implies reduced crew, fuel, berthing, insurance and maintenance costs. The only remaining constraints on ship size are the capacity of ports, harbors, and canals to accommodate them.

4. Speed. The average speed of ships is about 15 knots, which is 28 km per hour. Under such circumstances, a ship would travel about 575 km per day. More recent ships can

travel at speeds between 25 to 30 knots (45 to 55 km per hour), but it is uncommon that a commercial ship will travel faster than 25 knots due to energy requirements. To cope with speed requirements, the **propulsion** and engine technology has improved from sails to steam, to **diesel**, to gas **turbines** and to nuclear (only for military ships; **civilian** attempts were abandoned in the early 1980s). Since the invention of the **helix**, propulsion has improved considerably, notably by the usage of double helixes, but peaks were reached by the 1970s. Reaching higher marine speeds remains a challenge that is excessively costly to overcome. As a result, limited improvements in commercial marine speeds are foreseen. An emerging commercial practice, particularly in container shipping, concerns "slow steaming" where the operating speed is reduced to about 19–20 knots to reduce energy consumption.

5.　Specialization of ships. Economies of scales are often linked with specialization since many ships are designed to carry only one type of cargo. Both processes have considerably modified marine transportation. In time, ships became increasingly specialized to include general cargo ships, tankers, grain carriers, barges, mineral carriers, bulk carriers, Liquefied Natural Gas carriers, RO-RO ships and container ships.

6.　Ship design. Ship design has significantly improved from wood **hulls** to wood hulls with steel **armatures**, to steel hulls (the first were warships) and to steel, **aluminum**, and **composite** materials hulls. The hulls of today's ships are the result of considerable efforts to minimize energy consumption, construction costs and improve safety. Depending on its complexity, a ship can take between 4 months (container and crude carriers) and 1 year to build (cruise ship).

7.　Automation. Different automation technologies are possible including self-unloading ships, computer-assisted navigation (crew needs are reduced and safety is increased) and global positioning systems. The general outcome of automation has been smaller crews being required to operate larger ships.

8.　The shipping industry has a very international character, particularly in terms of ownership and flagging. The ownership of ships is very broad. While a ship may be owned by a Greek family or a Japanese corporation, it may be flagged under another nationality. There are two types of registers, national registers, and open registers, which are often labeled as "flags of convenience". Using flags of convenience allows ship owners to obtain lower registration fees, lower operating costs, and fewer restrictions. The marine industry is now more **deregulated** than before because of technical changes, mainly containerization and open registry ships operating under **fiscal** shelters. As of 2017, about 70% of the global tonnage was registered under a flag of convenience, with Panama and Liberia being the most **prevalent**. The marine shipping industry offers two major types of services:

9. Charter services (also known as Tramp). In this form of service, a maritime company rents a ship for a specific purpose, commonly between a specific port of origin and destination. This type of shipping service is notably used in the case of bulk cargo, such as petroleum, iron ore, grain or coal, often requiring specialized cargo ships that become the load unit (the whole contents of the ship are usually traded).

10. Liner shipping services. It involves a regularly scheduled shipping service often calling several ports along an inter-range route. The emergence of post-Panamax containerships has favored the setting of inter-range services since the marine landbridge of Panama is no longer accessible to this new class of ships. To ensure schedule reliability, which rarely exceeds 50%, frequency and a specific level of service (in terms of port calls), many ships can be allocated to a single route, which can take different shapes.

11. An important historic feature of oceanic liner transport is the operation of "conferences". These are formal agreements between companies engaged on particular trading routes. They fix the rates charged by the individual lines, operating for example between Northern Europe and the East Coast of North America. Over the years in excess of 100 such conference arrangements have been established. While they may be seen as anti-competitive, the conference system has always escaped **prosecution** from national anti-trust agencies. This is because they are seen as a mechanism to stabilize rates in an industry that is **inherently** unstable, with significant variations in the supply of ship capacity and market demand. By fixing rates exporters are given protection from swings in prices and are guaranteed a regular level of service provision. Firms compete on the basis of service provision rather than price.

12. A new form of inter-firm organization has emerged in the container shipping industry since the mid-1990s. Because of the costs of providing ship capacity to more and more markets are escalating beyond the means of many carriers, many of the largest shipping lines have come together by forming strategic **alliances** with former competitors. They offer joint services by pooling vessels on the main commercial routes. In this way, they are each able to commit fewer ships to a particular service route and deploy the extra ships on other routes that are maintained outside the alliance. The alliance services are marketed separately, but operationally involve close cooperation in ports of call selection and in establishing schedules. The alliance structure has led to significant developments in route **alignments** and the economies of scale of container shipping. The consequences have been a concentration of ownership, particularly in container shipping, the level of which is causing concerns among various na-

tional regulatory bodies that see such developments as potentially unfair competitive practices.

13. Carriers have the responsibility to establish and maintain profitable routes in a competitive environment. This involves three major decisions about how such a marine network takes shape:

14. Frequency of service. Frequency is linked with more timely services since the same port will be called more often. A weekly call is considered to be the minimum level of service but since a growing share of production is time-dependent, there is pressure from customers to have a higher frequency of service. A trade-off between the frequency and the capacity of service is commonly observed. This trade-off is often **mitigated** on routes that service significant markets since larger ships can be used with the benefits of economies of scale.

15. Fleet and vessel size. Due to the basic marine economics, large ships, such as post-Panamax containerships, offer significant advantages over long distances. Shipping lines will obviously try to use this advantage over their long-distance routes, keeping smaller ships for feeder services. In addition, a large enough number of ships must be allocated to ensure a good frequency of service. To keep their operations consistent, shippers also try to have ships a similar size along their long-distance inter-range routes. This is not an easy undertaking since economies of scale force the introduction of ever-larger ships which cannot be added all at once due to extensive financial requirements and the capacity of shipbuilders to provide them. So, each time a bigger ship is introduced on a regular route, the distribution system must adapt to this change in capacity.

16. Number of port calls. A route that involves fewer port calls is likely to have lower average transit times in addition to requiring a smaller number of ships. Conversely, too few port calls could involve difficulties for the cargo to reach inland destinations remote from the serviced ports. This implies additional delays and potentially the loss of customers. An appropriate selection of port calls along a marine **facade** will help ensure access to a vast commercial hinterland.

17. The global marine transportation system has substantially evolved to form networks within networks, connecting systems of circulation and enabling global trade. Without marine shipping, globalization could not have taken place to such an extent.

(1,592 words)

➤ *New Words*

scrap [skræp]	*v.*	to dispose of（something useless or old）废弃;使解体;拆毁

Of the above-mentioned measures, the most important is to scrap more old vessels as quickly as possible.

以上措施中,最重要的一项就是尽快拆解更多老旧船。

drawback [ˈdrɔːbæk]	*n.*	a disadvantage or problem that makes something less attractive 缺点,不利条件

A drawback is that LCDs consume a lot of power because they are lit from behind.

液晶显示器的一个缺点是会消耗大量电能,因为它们是从后面发光的。

propulsion [prəˈpʌlʃn]	*n.*	a propelling force 推进;推进力

Liquified natural gas will increasingly be used for long-haul truck, train and ship propulsion.

长途卡车、火车和船舶将越来越多地使用液化天然气提供动力。

diesel [ˈdiːzl]	*n.*	a heavy mineral oil used as fuel in diesel engines 柴油;柴油机

In spite of huge gas reserves, Libya's power plants mostly run on diesel fuel.

尽管利比亚天然气储量巨大,但该国的发电厂大多使用柴油。

turbine [ˈtɜːbaɪn]	*n.*	a machine or an engine that receives its power from a wheel that is turned by the pressure of water, air or gas 涡轮机;涡轮

Turbine manufacture and installation are also set to become major sources of employment, with one trade body predicting that the sector will generate 2 million jobs worldwide by 2020.

涡轮机制造和安装行业也会成为就业岗位的主要来源,一家贸易机构预计,到 2020 年,该行业将在全球创造 200 万个就业岗位。

civilian [səˈvɪliən]	*a.*	associated with or performed by nonmilitary citizens as contrasted with the military 民用的,百姓的,平民的

If the United States can loosen its export controls on civilian high-tech products, its export competitiveness can be further improved.

如果美国能够放宽对民用高科技产品的出口限制,将大大提升其出口竞争优势。

helix [ˈhiːlɪks]	*n.*	a spiral shape or form 螺旋,螺旋状物

DNA is a double-stranded helix made from four different types of sub-molecule.

脱氧核糖核酸是由四种不同类型的亚分子组成的双链螺旋。

hull [hʌl]   *n.* the main, bottom part of a ship 船体;船身;(果实的)外壳

The impact buckled the Titanic's hull and sent sea water pouring into six of its supposedly watertight compartments.

撞击使得"泰坦尼克号"的船体发生弯曲,海水随之涌进了六个水密舱室。

armature ['ɑːmətʃə(r)]   *n.* the protective covering of an animal or plant 保护层;盔甲;电枢

They set out for the adventure in a wooden ship with steel armatures.

他们驾着一艘铁甲木船开始了海上探险。

aluminum [ə'luːmɪnəm]   *n.* a silvery ductile metallic element found primarily in bauxite 铝

Despite an industry-wide consensus that aluminum prices would fall in 2019, the opposite has happened.

尽管业界普遍认为铝价将在 2019 年下跌,但事实却恰恰相反。

composite ['kɒmpəzɪt]   *a.* consisting of separate interconnected parts 复合的,合成的

The airlines bought it—even though the technical hurdles to building a carbon-composite airliner were formidable.

尽管制造碳复合材料飞机面临很多技术难题,但航空公司还是买账了。

deregulate [ˌdiː'regjuleɪt]   *v.* to lift the regulations on 解除对……的管制;撤销对……的规定或限制

In 1998, most European Union countries deregulated their domestic telecoms markets, allowing upstarts to compete with national monopolies.

1998 年,大多数欧盟国家计划解除了对国内电信市场的管制,允许新兴企业与国家垄断企业进行竞争。

fiscal ['fɪskl]   *a.* connected with government or public money, especially taxes 会计的,财政的;国库的

The effectiveness of fiscal easing depends on its composition as well as its size.

财政放宽能否有效,取决于其内容,也取决于其规模。

prevalent ['prevələnt]   *a.* that exists or is very common at a particular time or in a particular place 普遍的,广传的;流行的

Liver cancer is far more prevalent overseas, with 600,000 new cases a year worldwide.

肝癌在海外更为普遍,全球每年有 60 万例新发病例。

prosecution [ˌprɒsɪ'kjuːʃn]   *n.* the institution and conduct of legal proceedings against a defendant for criminal behavior 起诉,检举;进行;经营

Under international law, diplomats living in foreign countries are exempt from criminal prosecution.

根据国际法,驻外外交官员可免受刑事起诉。

inherently [ɪnˈherəntlɪ]   *ad.* cannot be removed easily 天性地;内在地;固有地

Experts point out that it doesn't imply that men are inherently less capable of raising children while statistics may favor women as parents.

有专家指出,虽然统计数据表明女性更适合养育孩子,但这并不意味着男性在这方面天生就差。

alliance [əˈlaɪəns]   *n.* a group of countries or political parties that are formally united and working together 联盟,联合;联姻

The President's visit was intended to cement the alliance between the two countries.

总统来访,是为了加强两国的联盟。

alignment [əˈlaɪnmənt]   *n.* support for a particular group, especially in politics, or for a side in a quarrel or struggle 结盟;队列,成直线;校准

Only 14 percent of companies globally have strong alignment between their innovation strategy and business strategy.

全球只有14%的公司在创新战略和商业战略之间达成了稳固结盟。

mitigate [ˈmɪtɪgeɪt]   *v.* to make something less unpleasant, serious, or painful 减轻;缓和

Efforts are under way to mitigate the effects of overpopulation and climate change.

人们正在努力减轻人口过剩和气候变化带来的负面影响。

facade [fəˈsɑːd]   *n.* the face or front of something 外观;正面;表面

The cathedral was eventually completed in 1490, though the Gothic facade remains unfinished.

那座大教堂最终于1490年建成,但其哥特式风格的外观一直未能完工。

➤ *Phrases and Expressions*

aim at 针对;瞄准;目的在于
depend on 取决于;依赖;依靠
be labeled as 被标记为;被贴上……的标签
engage on 从事;开始
be seen as 被看作;被视为
on the basis of 根据;基于……

to an extent 在一定程度上;到达……程度

➤ *Terminology*

slow steaming 减速航行,是集装箱航运公司为了节能减排和降低营运成本所应用的主要手段,被认为是目前航运业实现减排目标最能够接受的措施。在业界看来,减速航行不仅可节省大笔运营费用,还能有效地调节当前市场供需不平衡的状态,增加企业收益。由于部分船东一致看好通过降低油耗量来降低各项排放的方式,减速航行成为业界应对"限硫令"的选择之一。

global positioning system 全球定位系统

flag 船旗

national register 国家登记

open register 开放登记

flags of convenience 方便船旗,简称方便旗,一种船舶登记制度,也称巴利洪登记制,因巴拿马、利比亚、洪都拉斯三国最先实行开放性船舶登记制而得名,指一国准许他国所拥有或经营的船舶在该国登记并悬挂该国国旗航行的制度,所悬国旗作为方便船旗。有些商船为了逃避本国的法令管制、减少税收的交纳或工资等费用的支出等,选择在别国注册并悬挂该国旗帜。国际上已出现抵制方便船旗的趋势。《联合国海洋法公约》规定"国家和船舶之间必须有真正联系",以防止这种方便船旗的做法。

charter service 租船服务

liner shipping service 班轮运输服务

landbridge 本文中指海底大陆架,通常情况指大陆桥,是连接两个海洋之间的陆上通道,是横贯大陆的、以铁路为骨干的、避开海上绕道运输的便捷运输大通道,主要功能是便于开展海陆联运,缩短运输里程。

conference 班轮公会,又称航运公会,也称为水脚公会,是国际海上货物运输中,主要海运国家经营班轮运输的航运公司为保护和协调彼此间的权益而组成的超国家航运垄断组织。19 世纪末,国际航运竞争日益激烈,为避免因竞争跌价争揽货源而损害各自的利益,1875 年,7 家英国航运公司组成联合王国—加尔各答班轮公会。协议规定各自的船舶发航艘次和最低运价。此后班轮公会有很大发展,目前全世界已有 360 多个班轮公会,遍布各主要班轮航线,由海运发达国家航运公司控制。目前世界航运市场有 4 种运输方式:班轮公会运输、非班轮公会运输、无船承运人运输、不定期船运输。

anti-trust agency 反垄断机构

national regulatory body 国家监管机构

➤ *Proper Names*

Liberia 利比里亚

Panamax 巴拿马型船,为适合巴拿马运河船闸专门设计的一种大型船只,建造时精确匹配巴拿马运河船闸的限制船宽和吃水,以便在适应巴拿马运河航道的前提下运送尽可能多的货物。

➤ *Translation*

1. These drawbacks are particularly constraining where goods have to...（Para. 1）

   尤其是需要短途运输货物,或者托运人需要尽早交货的时候,上述缺点就更加明显。

2. For ship owners, the rationale for larger ships implies reduced crew...（Para. 3）

   对于船东而言,船越大(载货越多,单次载货多了,出航次数就可以相应减少),人力成本就越低,油料消耗也会降低,泊位开支、保险支出和维修成本都会相应减少。

   【首先,根据上下文增译"载货越多,单次载货多了,出航次数就可以相应减少";其次,根据汉语板块式结构的特点,将形容词比较级"larger"作为表语使用,从而扩充成为一个分句;另外,将来源于动词的形容词"reduced"还原译为动词,并根据不同搭配译为"减少"或者"降低"。】

3. An emerging commercial practice, particularly in container shipping...（Para. 4）

   尤其在集装箱运输行业,一种运营手段逐渐浮出水面,这就是"减速航行",即将航行速度降低到大约 19 至 20 节,以便降低能耗。

4. The hulls of today's ships are the result of considerable efforts...（Para. 6）

   现在的船体材料,在选材上主要考虑的是尽量减少能耗,降低造船成本,提高安全性。

5. The general outcome of automation has been smaller crews being...（Para. 7）

   总体来说,自动化的结果就是船虽然比以前大了,但是需要的船员却比以前少了。

6. Using flags of convenience allows ship owners to obtain lower...（Para. 8）

   对于船东来说,使用方便船旗注册费用低,运营成本小,操作限制少。

7. In this form of service, a maritime company rents a ship for a specific...（Para. 9）

   这种服务形式,通常是海事公司为了某一特定目的租用船只,在某特定始发港和目的港之间来回运行。

8. The emergence of post-Panamax containerships has favored...（Para. 10）

   受大陆架水深的限制,超巴拿马型船(不受巴拿马运河尺寸限制的大型集装箱船)无法通过巴拿马运河(只能选择跨海域航线),因此,这种船出现以后,跨海域航线就受到了青睐。

   【首先,根据英语句子先中心后外围、汉语先外围后中心呈现信息的特点,采取逆序翻译;其次,增译注释内容"(不受巴拿马运河尺寸限制的大型集装箱船)"补充背景信息,增译"只能选择跨海域航线"使译文逻辑通顺;另外,考虑英语常有名词、动词、形容词之间的词性转换,翻译时可根据上下文以及汉语表达习惯选择合适的词性,将名词"emergence"转译为动词"出现",将动词"favor"转译为名词"青睐",将形容词"accessible"转译为动词"通过",同时考虑到汉语多短句、多分句的特点,将主句主语从原文中拆分出来,单独成句。】

9. The consequences have been a concentration of ownership...（Para. 12）

   这就造成了所有权过度集中的现象,特别是在集装箱船运输行业,这种现象尤为突出,已经引起了各国监管机构的关注,他们认为如此发展下去,有可能会引起不公平竞争。

10. This is not an easy undertaking since economies of scale force... (Para. 15)

这不是一件容易的事,要实现规模经济,就必须投入越来越大的船只,可是那么多的大船,不是一天就能建好的,不单是财力所限,造船厂也没有这么大的产能。

11. Conversely, too few port calls could involve difficulties for... (Para. 16)

相反,沿途停靠港口过少的情况下,如果货物目的地是远离经停港口的内陆,那就不是很方便了。

 **Exercises**

1. Fill in the blanks with the proper given words, and then translate the sentences into Chinese.

aim at	maintenance	civilian	abandon
foresee	complexity	be labeled as	restriction
prevalent	engage on	competitive	prosecution
inherently	alliance	consistent	to an extent

1) All the parties have realized the _____ of the Peninsula nuclear issue, whose resolution requires prolonged efforts.

2) The authorities offered to stop firing missiles if the rebels agreed to stop attacking _____ targets.

3) We are elevating environmental issues in our diplomatic relationships and forging new partnerships to better _____ those global challenges.

4) Cooperation should _____ pursuit of mutual beneficial win-win results and common development.

5) Although she refuses to _____ a feminist, her novels embody her feminist thought to the fullest.

6) Signs of slower growth have, _____, been welcomed by investors hoping that they could pave the way for a relaxation of monetary policy and a launch of fresh fiscal stimulus.

7) We can _____ that new technologies might be developed in the coming century that would further expand our power over nature and ourselves.

8) The school pays for heating and the _____ of all the classroom buildings according to the contract.

9) They did not persuade him to _____ the war but the situation did force him to reappraise his strategy.

10) The level of the administration inevitably has _____ and influence on the social construction and development.

11）Each member of the _____ agrees to take such action as it deems necessary, including the use of armed force.

12）We know that this is pretty _____ in our culture, but we don't know much about it from a scientific standpoint.

13）When it comes to law enforcement, varying jurisdictions and laws complicate the _____ of cybercriminals.

14）As a(an) _____ social species, we are constantly reinforcing our social connections through communication.

15）The US strategy on this issue is not, at least for the moment, _____ with strategies elsewhere.

16）GATT member countries have largely agreed to replace inefficient protectionist policies with a _____ free market system for world trade.

2. Translate the following passage into Chinese.

An important historic feature of oceanic liner transport is the operation of "conferences". These are formal agreements between companies engaged on particular trading routes. They fix the rates charged by the individual lines, operating for example between Northern Europe and the East Coast of North America. Over the years in excess of 100 such conference arrangements have been established. While they may be seen as anti-competitive, the conference system has always escaped prosecution from national anti-trust agencies. This is because they are seen as a mechanism to stabilize rates in an industry that is inherently unstable, with significant variations in the supply of ship capacity and market demand. By fixing rates exporters are given protection from swings in prices and are guaranteed a regular level of service provision. Firms compete on the basis of service provision rather than price.

3. Translate the following passage into English.

全球最大的集装箱运输公司马士基集团日前表示，去年年末爆发的新冠病毒导致今年年初开始航运需求大幅下滑，因此公司不得不取消 50 余条往返亚洲的航线。随着疫情蔓延，不仅人们的生活受到了影响，企业也受到了冲击，甚至可能使经济陷入瘫痪。仅就航运业来说，我们的日常生活用品有 80% 都要通过海洋运输，因此，对新冠病毒感到忧心忡忡的其实并不仅仅是卫生官员，还有航运及其相关企业，他们也希望疫情早日到达拐点，尽快得到缓解。然而，截至目前，还没有好转的迹象，形势依然还很渺茫。

Chapter 7

Marine Information Engineering

Unit 1

Marine Information Data

Progress in Satellite Remote Sensing

1. The surface exchange of heat and **momentum** is responsible for the dynamic circulation of the atmosphere and oceans. Heat **fluxes** in the surface layers of the ocean result from two opposing processes: that of direct heating by the sun, which tends to promote vertical **stratification**, and that of mixing by wind, which tends to erode stratification. For the purpose of modelling heat flux, the solar irradiance is usually treated as the sum of two components: a near-infrared component, whose path length in the ocean is only a few millimetres (the nonpenetrating part), and a component with a much longer path length (the penetrating part), covering approximately the visible **spectrum**. Energetically, the two components are roughly equal in magnitude.

2. The depth of the mixed layer affects the potential growth rate of **phytoplankton** in the layer. In a shallow layer, through which phytoplankton cells are mixed, the average light intensity experienced by the cells will be greater than in a deeper mixed layer. Phytoplankton growth occurs through **photosynthesis**, which is a process forced by light: all other things being equal, growth conditions in a shallow mixed layer will be more favourable than those in a deeper mixed layer. The possibility therefore exists for a positive feedback between increasing **chlorophyll** concentration and increasing SST (sea surface temperature), a mechanism that is more likely to occur in spring in temperate latitudes, in association with the **vernal** bloom of phytoplankton. It can modify the seasonality (**phenology**) of phytoplankton in these latitudes. Wu showed that, as well as advancing the phenology, the influence of phytoplankton on the seasonal progression of mixed-layer temperature and depth was such as to increase the maximum SST for the year by 1.5 ℃ in the Labrador Sea, and 2.5 ℃ over the Grand Bank of Newfoundland. Gnanadesikan studied the impact of bio-optical heating on **cyclone** activity in the tropical Pacific.

3. Horizontal gradients in the chlorophyll field can also lead to modification of both the horizontal circulation and the vertical circulation. Fronts in the chlorophyll field are commonly seen in ocean colour imagery. Physically, there will be an imbalance in solar heating rates across the front: the biomassrich side will be heated more at the surface than the low-biomass side, resulting in a density discontinuity across the front, **perturbing** the frontal circulation. The horizontal **velocities** along the front are

2 cm/s; therefore, the vertical velocities are roughly 0.2 mm/s, which is potentially significant for re-supply of nutrients at the front.

4. In brief, there is considerable evidence that variability in the optical properties of the ocean has an impact on the distribution of solar induced heating of the ocean. In the open ocean, optical variability is determined largely by variability in phytoplankton and associated material, such as coloured dissolved organic matter and coloured **detrital** matter. However, such effects are yet to be included in a routine manner into models of ocean heat fluxes or into studies of biological—physical coupling in Earth system models, though such models are now emerging and they are being incorporated increasingly into biogeochemical and ecosystem models of the ocean. Further work in this area is likely to have major effects on our capability to predict how global circulation may change in the future.

5. Many of the research areas and applications presented in this review would benefit from the provision of long term datasets. As such, the European Union Copernicus programme (formerly known as the Global Monitoring for Environment and Security, GMES, programme) offers a huge opportunity. This programme has started and will provide a long term (15 years) monitoring capability for the marine environment. The space component of this programme is formed by a fleet of satellites called the **Sentinels**, most of which are designed to monitor different aspects of the oceans and atmosphere. For instance, Sentinel 1 carries an advanced SAR instrument and Sentinel 3 carries a radar **altimeter** and visible and thermal **infrared** instruments.

6. The development of the Global Navigational Satellite Systems-Reflectrometry (GNSS-R) promises to provide advances in wind speed and wave monitoring. GNSS-R is a relatively new concept that exploits existing data from GNSS systems including the European Galileo and US Global Positioning System (GPS) systems by turning them into a bistatic radar system. The main advantages of GNSS-R lie in the global availability of GNSS signals, which provides a dense coverage over the Earth and offers the potential for significant improvements in **spatio-temporal** sampling of ocean winds and waves. The receiver needed to capture the surface reflections is small and cheap, low-weight/low-power and can be accommodated easily on small satellites, thus reducing the cost of a mission and allowing for **constellations** of multiple GNSS-R receivers. The GNSS-R capability is already present on some satellites, including the UK Disaster Monitoring Constellation (DMC) satellite.

7. National Aeronautics and Space Administration (NASA) has recently launched two remote sensing instruments to be mounted on the International Space Station (ISS). The ISS is also a satellite, and so provides a unique low Earth orbit platform on which to mount sensors. Its orbit pattern (6 orbits a day over **equatorial** to mid latitudes) provides a unique viewing position for cyclones and hurricanes. The ISS already carries a visible spectrum sensor (the hyper-spectral imager for the coastal oceans, HICO) and

the ISS-RapidSCAT, which is a scatterometer designed to provide data for climate research, weather predictions and hurricane monitoring. A further instrument planned for the ISS, the Cloud-Aerosol Transport System (CATS), may also prove useful for oceanography, if only to improve the quantification of atmospheric effects on other types of remotely sensed oceanographic data.

8. The creation and exploitation of miniature satellite constellations (e. g. femto-, pico-, nano-, micro- and mini satellites), such as the programme being developed and released by the US Planet Labs start-up company, are likely to provide the scientific and operational communities with new datasets suitable for studying and monitoring physical oceanography. The launch of collections of miniature satellites (i. e. more than 20 in orbit at any one time) could enable sub-monthly processes to be monitored and characterized. The Planet Labs satellites are being developed to carry simple optical camera systems, and early data releases from these satellites suggest that the data could be exploited to study surface ocean currents, sea ice and heat fluxes using some of the methods described in Sections II, III and VI of this review.

9. Our ability to monitor the physical characteristics and features of our oceans is imperative to help us understand and predict our climate, to support the search and provision of energy and to support sustainable food harvesting and production. In this paper we have provided an overview of how the different remote sensing techniques and approaches can be used to observe the physical properties of our oceans. We have then critically reviewed recent advances in how these observations are being used to understand, **quantify** and monitor surface currents, storm surges, sea ice, atmosphere—ocean gas fluxes and surface heat fluxes. We highlight that satellite remote sensing clearly has a significant role to play in these scientific areas by providing large spatial scale observations and measurements. However, satellite remote sensing cannot be used in isolation, so the use of in situ and model data in conjunction with satellite remote sensing data is key to enabling the full exploitation of these space based observations. The current drive to establish long term monitoring capabilities through commercial enterprises and government (agencies and policy) will **underpin** our future ability to provide regular and routine monitoring of our oceans. Novel platform and technology developments, including GNSS-R, commercial nanosatellites and the International Space Station will continue to provide us with new perspectives and ways of monitoring our oceans.

(1,282 words)

➢ *New Words*

momentum [məˈmentəm]   *n.* the product of a body's mass and its velocity 动力
The spinning top was losing momentum.
旋转的陀螺正在失去动力。

flux [flʌks]   *n.* continuous movement and change 不断的变动;不停的变化
Both quantum mechanics and chaos theory suggest a world constantly in flux.
量子力学和混沌理论都指出世界处在不断变化中。

stratification [ˌstrætɪfɪˈkeɪʃn]   *n.* a layered configuration 层理;成层
Physical differences between regions, such as currents and the stratification of the water column, will also complicate the response of food webs.
在不同区域,水的物理性质(如水体的流动与分层)也会有所差异,这也将使鱼类对食物网的反应复杂化。

spectrum [ˈspektrəm]   *n.* an ordered array of the components of an emission or wave 光谱;频谱
Matter between the two fills in the spectrum with colours such as green.
介于两者之间的物质会填满诸如绿色在内的整张光谱。

phytoplankton [ˈfaɪtəʊˌplæŋktən]   *n.* photosynthetic or plant constituent of plankton; mainly unicellular algae 浮游植物(群落)
The researchers found that the most notable phytoplankton declined in waters near the poles and in the tropics, as well as the open ocean.
研究人员发现,在两极附近的水域和热带地区的浮游植物的跌幅最为显著,在开阔的海洋也是如此。

photosynthesis [ˌfəʊtəʊˈsɪnθəsɪs]   *n.* synthesis of compounds with the aid of radiant energy (especially in plants) 光合作用
During photosynthesis, leaves have to produce chlorophyll to capture the energy from the sun, and it is this chlorophyll that gives them their green colour.
在光合作用期间,各种叶子不得不产生叶绿素来捕获太阳能,而就是这种叶绿素给了各种叶子绿色。

chlorophyll [ˈklɒrəfɪl]   *n.* any of a group of green pigments found in photosynthetic organisms; there are four naturally occurring forms 叶绿素

Scientists use these changes in ocean color to estimate chlorophyll concentration and the biomass of phytoplankton in the ocean.

科学家利用海洋颜色的变化来估计叶绿素含量以及海洋浮游植物的生物量。

vernal ['vɜːnl]   *a.* of or characteristic of or occurring in spring; suggestive of youth; vigorous and fresh 春天的

The vernal equinox occurs when the center of the sun crosses the Equator.

当太阳中心直射点在赤道之时,便是春分。

phenology [fəˈnɒlədʒɪ]   *n.* the study of periodic plant and animal life cycle events and how these are influenced by seasonal and interannual variations in climate, as well as habitat factors (such as elevation) 生物气候学,物候学

Biological and environmental factors are the main factors that influence plant phenology, especially the latter, among which the dominating factors are air temperature, photoperiod and water.

影响植物物候学变化的因子主要有生物因素和环境因素,后者对物候学影响更显著,其中气温、光照和水分是最主要的影响因子。

cyclone ['saɪkləʊn]   *n.* (meteorology) rapid inward circulation of air masses about a low-pressure center; circling counterclockwise in the northern hemisphere and clockwise in the southern 气旋

This Great Red Spot is still present in Jupiter's atmosphere, more than 300 years later. It is now known that it is a vast storm, spinning like a cyclone.

这个大红斑在木星大气层已经存在了超过 300 年,现在我们知道,大红斑是一个巨大的风暴,它像气旋一样在不断地旋转中。

perturb [pəˈtɜːb]   *v.* to throw into great confusion or disorder 扰乱

Stellar pathways can perturb the orbits of comets.

行星的运行会扰动彗星的轨道。

velocity [vəˈlɒsətɪ]   *n.* distance travelled per unit time 速度

So notice that the acceleration is constant in time, which is not changing but the velocity is changing.

请注意此处的加速度是一个常数,不随时间变化而变化,而速度是变化的。

detrital [dɪ'traɪtl]   *a.* formed from debris 碎屑的;由岩屑形成的

Sodium carbonate deposits may be classified into three major types, Quaternary salt lake, Quaternary sand-covered playa and Tertiary continental detrital rock, all three having been found in China.

碳酸钠矿床有三大类型:第四纪盐湖型、第四纪砂下湖型及第三纪陆相碎屑岩型,在中国均有发现。

sentinel ['sentɪnl]   *n.* a person employed to watch for something to happen 哨兵

My mother tiptoed from their room to find me patiently waiting in the same spot she had left me, standing like a sentinel in the hallway.

母亲从她们房间蹑手蹑脚地走出来,看到我耐心地等在原地,像个哨兵似的站在走廊上。

altimeter ['æltɪmiːtə(r)]   *n.* an instrument that measures the height above ground; used in navigation 测高仪,高度计

The new probe was built by using improved electronics and batteries, and an extra radar altimeter, a device that will fire microwaves at the Arctic and Antarctic ice to reveal its thickness.

新的探测器在建造时使用了改进的电子设备和电池组,并且添加了新的雷达高度计,这是一种发射微波测定两极冰层厚度的装置。

infrared [ˌɪnfrə'red]   *a.* having or employing wavelengths longer than light but shorter than radio waves; lying outside the visible spectrum at its red end 红外线的;(设备、技术)使用红外线的;对红外辐射敏感的

Using the Hubble Space Telescope, a team of US and European researchers have captured infrared images of a blurry object they believe to be the oldest galaxy ever seen by man.

一组美国和欧洲的研究人员利用哈勃太空望远镜拍取了一个物体的模糊红外线图像,他们认为这是人类所观测到的最古老的星系。

spatio-temporal [ˌspeɪʃɪəʊ'tempərəl]   *a.* of or relating to space and time together (having both spatial extension and temporal duration) 时空的

Spatio-temporal database management system (STD-

BMS) is a kind of DBMS that can deal with both spatial and temporal data.

时空数据库管理系统是能够同时处理时态数据和空间数据的数据库管理系统。

constellation [ˌkɒnstəˈleɪʃn]　　　*n.* a configuration of stars as seen from the earth 星座;星群

This planet sits in the southern constellation of Fornax.

这颗行星位于天炉星座的南部。

equatorial [ˌekwəˈtɔːriəl]　　　*a.* of or relating to or at an equator 赤道的,近赤道的

Thies says that this could explain the origin of some of the exoplanets that have been detected with orbits significantly tilted with respect to their star's equatorial plane.

蒂斯说,"它可以解释某些外行星的起源,这些外行星已被检测出它们的运行轨道明显地与它们所围绕恒星的赤道平面相倾斜"。

quantify [ˈkwɒntɪfaɪ]　　　*v.* to express as a number or measure or quantity 量化

Its ability to destroy ozone has been known for decades, but the new research is the first to quantify the danger and compare it to other gases.

几十年前人们就知道它能破坏臭氧层,但这项新的研究首次量化了这种危险的程度,并与其他气体做了比较。

underpin [ˌʌndəˈpɪn]　　　*v.* to support with evidence or authority or make more certain or confirm; support from beneath 巩固

In addition, we will ask all developed countries to reduce emissions by at least 80% in the same period and underpin these efforts through robust and comparable mid-term reductions.

此外,我们将要求所有发达国家在同一时期减少80%的排放,通过在中期强有力和类似的努力加以巩固。

➤ *Phrases and Expressions*

in association with 与……相联系;与……联合
at the front 在前面;在前线
have an impact on 对……有影响;对……造成冲击

be likely to 倾向于,很有可能
as such 同样地;本身;就其本身而论
lie in 在于……;睡懒觉;待产
allow for 考虑到,虑及
if only 只要;要是……多好
in conjunction with 连同,共同;与……协力
be key to doing 是做……的关键

➤ *Terminology*

Earth system model 地球系统模式
bistatic radar system 双基雷达系统

➤ *Proper Names*

Labrador Sea 拉布拉多海(位于格陵兰与拉布拉多岛之间)
Grand Bank of Newfoundland 纽芬兰大浅滩
Global Monitoring for Environment and Security(GMES) 全球环境与安全监测系统
Global Navigational Satellite Systems-Reflectrometry(GNSS-R)全球导航卫星系统-反射测量
Disaster Monitoring Constellation(DMC)灾害监测星座
National Aeronautics and Space Administration(NASA)美国国家航空航天局
International Space Station(ISS)国际空间站
the hyper-spectral imager for the coastal oceans(HICO)沿海海洋的高光谱成像仪
ISS-RapidSCAT 国际空间站的快速散射计
Cloud-Aerosol Transport System(CATS)云-气溶胶运输系统

➤ *Translation*

1. Heat fluxes in the surface layers of... and that of mixing by wind, which tends to... (Para. 1)
   海洋表层的热通量是由两种相反的过程引起的:一种是太阳的直接加热,它倾向于促进垂直分层;另一种是风的混合,它倾向于侵蚀分层。
   【这个句子的冒号后面由两个 that of 引导的含有非限定性定语从句的并列结构构成,翻译时如果直译为"那类……的,其倾向于……",译文会令人费解,达不到忠实于原文的目的。这时,我们可考虑采用换形译法的翻译方法,译为"一种是……它倾向于……,另一种是……它倾向于……"。这和汉语的语言习惯相关,译文更为地道。英语多用 that 替代上文内容,在进行翻译时需注意换形而不是换义。】

2. In a shallow layer, through which phytoplankton cells are mixed... greater than in a deeper mixed layer. (Para. 2)
   在浮游植物细胞混合的浅层中,细胞所经历的平均光强度大于在较深的混合层中的光

强度。

3. The possibility therefore exists for a positive feedback..., in association with the vernal bloom of phytoplankton. (Para. 2)

因此,增加叶绿素浓度和增加海表温度之间存在正反馈的可能性,这一机制更可能发生在温带地区的春季,与浮游植物春季开花有关。

4. Horizontal gradients in the chlorophyll field... and the vertical circulation. (Para. 3)

叶绿素场的水平梯度也会导致水平环流和垂直环流的改变。

5. Physically, there will be an imbalance in solar heating rates... perturbing the frontal circulation. (Para. 3)

从物理上讲,正面的太阳能加热速率将是不平衡的:生物质面将比低生物质面在表面被加热得更多,导致正面的密度不连续,扰乱正面的循环。

(这里主要是指《自然地理学进展》杂志公布的海洋信息数据调查研究。)

6. In brief, there is considerable evidence that variability..., such as coloured dissolved organic matter and coloured detrital matter. (Para. 4)

简而言之,有相当多的证据表明,海洋光学性质的变化对太阳引起的海洋热的分布有影响。在开放的海洋中,光的可变性很大程度上是由浮游植物和相关物质的可变性决定的,如有颜色的溶解有机物和有颜色的碎屑物质。

7. The space component of this programme is formed by a fleet of satellites... the oceans and atmosphere. (Para. 5)

这一方案的空间部分是由一个称为"哨兵"的一组卫星组成,其中大多数卫星的设计目的是监测海洋和大气的各自不同方面。

8. The launch of collections of miniature satellites... to be monitored and characterized. (Para. 8)

发射小型卫星集合(即在同一时间内在轨道上的卫星超过 20 颗),可以使次月进程得到监测并探测其特征。

9. Our ability to monitor the physical characteristics and... to support sustainable food harvesting and production. (Para. 9)

我们监测海洋物理特征和特征的能力对于帮助我们了解并预测气候变化、搜寻能源供应,以及提供可持续的获取粮食与生产至关重要。

10. However, satellite remote sensing cannot be used in isolation, so the use of in.... (Para. 9)

然而,卫星遥感不能单独使用,因此,将现场和模型数据与卫星遥感数据结合使用是充分利用这些空间观测的关键。

【转性译法,即将译文语言的习惯进行词性转换。这个句子中 the use of ... 的"use"是名词,进行翻译时将其翻译为动词"使用",使译文能够被中国读者接受,从而使静态叙述转换成动态叙述。】

Exercises

1. Fill in the blanks with the proper given words, and then translate the sentences into Chinese.

surface	satellite	ocean	oceanography
flux	instrument	wave	reduce
space	launch	data	monitor
station	remote	release	current

1) He still had a lot of pent-up anger to _____.

2) However, as Mr. Lebrato and Dr. Jones report in *Limnology and* _____, when they analyzed devilfish tissues they found that the creatures were one-third carbon by weight.

3) However, as its name suggests, it runs parallel to the _____ of the material that is propagating it, rather than penetrating this material.

4) All inner scenes are shot inside this car. Tunnel views and _____ interiors will be later laid on the green background with the help of computer.

5) The plane is 107 feet from nose to tail, as big as a commercial jet seating 100 passengers, but with computers and _____ bays occupying most of the interior space and only a few windows.

6) We can then translate this information into estimates of the _____ diabetes and cardiovascular disease that can be attributed to the rise in consumption of these drinks.

7) Now we have a feeling for what the _____ elements look like. How do we manage them with our application?

8) To _____ your frustration with this process, it helps to know the six phases people go through whenever they are experiencing any type of change.

9) But who regulates the marketing, by _____ TV or Internet, of unhealthy lifestyles, including diets, tobacco products and pharmaceuticals of questionable quality?

10) This offloads some of the work that developers would otherwise need to do, such as writing scripts or bat files to _____ or configure an application.

11) You can show that if you put a detector on either slit the _____ probability function changes but that not explaining why it happens, it's just saying this is the way it does happen.

12) _____ looks quite different in the plane and in space because, in the plane, it is just another kind of line integral, while in space it is a surface integral.

13) Overfishing not only causes depletion in individual fish stocks, but also disrupts the entire ecosystems and food webs in the _____.

14) This procedure allows any queue on the device to be addressed through one definition, reducing the number of _____ queue definitions to the number of device queue managers.

15) But even if one day we manage to explain time and _____ in terms of something else, that only pushes the question to another level, because you then have to explain something else.

16) These systems are hardware or software devices that can help you _____ access across the network to and from your server as well as activity on the server itself.

2. Translate the following passage into Chinese.

The development of the Global Navigational Satellite Systems-Reflectometry (GNSS-R) promises to provide advances in wind speed and wave monitoring. GNSS-R is a relatively new concept that exploits existing data from GNSS systems including the European Galileo and US Global Positioning System (GPS) systems by turning them into a bistatic radar system. The main advantages of GNSS-R lie in the global availability of GNSS signals, which provides a dense coverage over the Earth and offers the potential for significant improvements in spatio-temporal sampling of ocean winds and waves. The receiver needed to capture the surface reflections is small and cheap, low-weight/low-power and can be accommodated easily on small satellites, thus reducing the cost of a mission and allowing for constellations of multiple GNSS-R receivers. The GNSS-R capability is already present on some satellites, including the UK Disaster Monitoring Constellation (DMC) satellite.

3. Translate the following passage into English.

迄今为止,所有的气体流量研究都集中在作为主要目标气体种类的二氧化碳上。对可溶气体进行参数化的研究能力也有助于对不可溶气体进行研究。例如,气体传递的非气泡成分可以使用可溶性气体数据(如数据库管理系统)进行校准,然后将结果应用于不溶性二氧化碳。自然界的循环系统也紧密相连(例如,氮循环和碳循环),因此,如果了解一氧化二氮的循环方式,可以支撑对二氧化碳数据的解释。甲烷和一氧化二氮都是气候学上的重要气体,都是水侧受控气体。因此,可以用方程来确定气体流量参数。

Unit 2
Model of Marine Information Data

Data Structures for NEMO Ocean Model

1. NEMO (Nucleus of the European Model of the Ocean) is one of the most widely used ocean models, of great significance to the European climate community. The NEMO model is written in **Fortran** 90 and its origins can be traced back over 20 years and, as such, the code base, **parallelized** using the message passing paradigm, has seen many computer architectures and programming environments come and go. The model is used in more than 240 projects in 27 countries for climate studies both at regional and global scales. **Portability**, longevity and computational performance are therefore highly important to the NEMO users and developers.

2. The NEMO model is parallelized using the MPI with a two-dimensional geographic domain decomposition. The initial domain described along the latitude, the longitude and the depth coordinates is spread to the parallel tasks dividing the horizontal grid in a regular **mesh**. Each task takes only one sub-domain made of a rectangular portion of the horizontal grid and all of the corresponding vertical levels (i. e. the sub-domain is a **parallelepiped**). Each sub-domain is enriched with a 1-point wide halo region that is used to exchange the data at the borders with the neighbouring tasks. The communication pattern is based on a 5-point **stencil** and uses blocking MPI communications.

3. This parallelization approach is common to many other ocean and fluid dynamics models since it is a natural way to exploit the fact that the same operations are performed at each point on the simulation grid.

4. *Outer loop parallelization.* The outer loop strategy (OMP1) is based on the parallelization of the **outermost** loop of any loop nest. Each computational loop will be parallelized by simply including the OMP DO statement before the outer loop. This means that all of the triply nested loops will be parallelized in the k (vertical/depth) dimension. The doubly nested loops, mainly dealing with the state variables defined on the sea surface and ocean floor, will be parallelized in the j (latitude) dimension. The resulting domain decomposition is depicted. Each thread takes the same horizontal domain assigned to the MPI process (of extent jpi $*$ jpj) and a portion of the vertical levels.

5. *D tiling parallelization.* The second approach (OMP2) is based on the idea of decomposing the MPI domain into a set of 3D tiles. Each tile is characterized by three pairs of **indices** that define the shape of the tile. The indices also specify the position of the tile within the MPI domain. Thus, each MPI process has to perform the operations for all the tiles belonging to its sub-domain. The OpenMP parallelization will then be introduced so as to simultaneously perform computation on the tiles assigned to an MPI process. This approach implies that during the initialization, each process must create its tiles according to a pre-defined domain decomposition strategy. The resulting domain decomposition is illustrated.

6. *Loops **merge** parallelization.* The last approach (OMP3) is based on the idea of flattening the triply and doubly nested loops such that they become single loops iterating over the total number of elements of the original, nested loop. In other words, it is possible to create a flat loop by merging together the nested loops using the OMP COLLAPSE directive.

7. The introduction of further levels of parallelization with OpenMP and the consequent **reshaping** of the subdomain assigned to each parallel thread has an impact on both the use of the memory hierarchy and the level of SIMD **vectorization**. For these reasons a version of NEMO with the array indices reordered could potentially make more efficient use of hardware capabilities in terms of cache use and instruction-level parallelism. Historically, NEMO and its ancestor, OPA, have been developed for vector architectures as opposed to the **scalar** architectures of current machines. As a consequence, three-dimensional or rank-three arrays in NEMO are declared with the k (depth) index last.

8. Recently, mini-applications, as lighter versions of complex high-performance computing (HPC) applications, have been used as a flexible test bed to facilitate software development. They can be used, as representative of the behaviour of the complete application, to understand the benefits deriving from a proposed software update without changing the entire code. The development process of a mini application starts from the identification of the bottlenecks of the real application. The computational **kernels** are extracted and a **stand-alone** application will be developed for each of them preserving the computational characteristics, the data structures and the communication pattern of the original model.

9. *Results.* Three kinds of test have been performed. The first set of tests aims to evalu-

ate the suitability of the **hybrid** parallelization on different architectures and to identify the **optimal** ratio of MPI processes to OpenMP threads on each. For these tests, the tra_adv_muscl kernel has been considered using two different domains: the first includes about 16 million grid-points and is similar to the Mediterranean Forecast System (MFS) configuration implemented at CMCC. The second domain is defined in order to **saturate** the available memory on the compute node of each architecture under test.

10. *Architecture comparison.* In this set of tests our aim is to provide some insight on the main multicore architectures available today in HPC **clusters** when running the pure OpenMP version of the most simple of the kernels, tra_ldf_iso. For this kernel we implemented only the outer-loop OpenMP approach since it has been demonstrated to be more performant. The comparison has been made between the following architectures: the AMD Magny Cours; IBM Power7; Intel Westmere, Intel Sandy Bridge; and Intel Xeon Phi (Knights Corner). In the experiments we executed the mini-application with only one MPI task while changing the number of threads.

11. The results reported demonstrate that the best absolute performance was obtained with 120 threads (i. e. two threads per core) on the Intel Xeon Phi. On more than 120 threads performance drops off rapidly in this case.

12. *Evaluation of array index reordering.* The last set of tests aims to evaluate the impact on performance of the array index reordering. For these tests the tra_adv_muscl kernel has been used, comparing the outer-loop and 2D tile-based parallel approaches. In contrast to the OMP1 and OMP3 approaches described earlier, here we collapse (flatten) only the outer two loops of a triply nested loop. For the 2D tiling, each MPI sub-domain is divided into tiles but only in the i, j (latitude–longitude) plane and these tiles are shared amongst OpenMP threads. Each tile retains the full domain extent in the third dimension. This decomposition can also be obtained using the OMP2 approach and slicing only along the longitudinal and latitudinal directions. This idea is illustrated where the MPI sub-domain is shown **decomposed** into eight tiles.

13. We have investigated a variety of approaches for introducing hybrid parallelism into the NEMO ocean model. Our experiments with individual, tracer-related kernels have shown that best absolute performance is obtained by simply parallelizing the outer loop with OpenMP. Good practice when considering thread-based parallelism, beyond the issues of load balancing and thread scheduling, concerns the optimal management of memory accesses and **cache** reuse. The best performance can be achieved when (i) the thread operates on a contiguous data set (an optimal domain decomposition can guarantee this property); (ii) the thread is bound with the core

using memory **affinity**; (ⅲ) false sharing is avoided (the mechanism whereby a cache line of a thread is invalidated by another thread which accesses different data which belong to the same cache line).

14. We have also shown that the change of array index ordering from k-last to k-first can provide a significant increase in performance in the strong-scaling limit when MPI sub-domains become relatively small. However, this gain comes at the cost of a massively intrusive and pervasive change to the NEMO code base. Since that code base is under continuous development, the benefits of such a change must be carefully considered.

15. The Intel Xeon Phi struggled to match the performance of the traditional Xeon CPU as soon as the kernel involved the thread **synchronization** and load imbalance introduced by halo-exchange (MPI) operations. The results also demonstrate that, for the traditional architectures, the management of multiple threads within a node is less efficient than the management of Linux processes and the pure MPI implementation is good when the sub-domain is relatively big. The hybrid implementation **outperforms** the pure MPI version once the sub-domain size becomes small on both the Intel and IBM BG/Q systems. However, on the IBM Power-based systems the pure MPI implementation was the most performant in all of our tests. Finally, considering all of the proposed approaches, we conclude that the introduction of OpenMP through the parallelization of the outer loop (over vertical levels) can be considered the best tradeoff between programmability, robustness and performance.

(1,461 words)

➤ *New Words*

Fortran ['fɔ:træn]

n. (abbr. = formula translator) a high-level programing language for mathematical and scientific purposes; stands for formula translation 公式翻译程式语言

Courses taken that would be useful for computer programming are: computer science, systems design and analysis, FORTRAN programming, PASCAL programming, operating systems, systems management.

对计算机编程有用的课程有:计算机学、系统设计与分析、公式翻译程式语言编程学、电脑语言编程学、操作系统、系统管理。

parallelize ['pærəlelaɪz]

v. to place parallel to one another 使程序(适合)进行计算

It uses all available processor to parallelize the operation. 它使用所有可用的处理器并行操作。

portability [ˌpɔ:təˈbɪlətɪ]

n. the quality of being light enough to be carried 可移植性

In order to achieve portability and extensibility, we would ideally like to defer inclusion of those details to a very late stage in development.

为了实现可移植性和可扩展性,理论上我们要推迟到开发得非常晚的阶段再加入那些细节。

mesh [meʃ]

n. the number of opening per inch of a screen; measures size of particles; contact by fitting together 网格

Follow this logic to its end and you understand why free software folks are advocating for net neutrality or trying to build open, alternative networks via mesh.

按照此逻辑,你就明白为什么自由软件人正在倡导网络中立性,或试图建立开放网络,通过网格技术的替代网络。

parallelepiped [ˌpærəleləˈpaɪped]

n. a prism whose bases are parallelograms 平行六面体

The general volume is a compact parallelepiped in the thickness of which the introduction of three patios bring light and allows a natural ventilation of the round local devices.

该建筑大体是一款紧凑型的平行六面体,三个露台可引进阳光,而其环绕的本地设备可引入自然通风。

stencil ['stensl]

n. device that has a sheet perforated with printing through which ink or paint can pass to create a printed pattern

模板

Each block was articulated using the existing vegetation as a stencil and building was thus carved out.

每栋大厦都使用现有的植被连接并作为一个模板,这样建筑就被雕刻出来。

outermost ['aʊtəməʊst]	a.	situated at the farthest possible point from a center 最外面的;最远的

Carbon monoxide has a simple but distinctive outermost orbital structure with two side-by-side lobes sticking out from the end of the tip, one with a positive phase and the other a negative phase.

一氧化碳有简单且独特的最外层轨道结构,两个并列的叶形从尖端的末梢伸出,一个是正相而另一个是负相。

indice ['ɪndɪs]	n.	index; trin(古法语)指数;标记体

It is deemed that peak values, especially the peak velocity, remain the most appropriate indice for destructiveness.

峰值特别是速度峰值仍然是地震破坏性的最合适指标。

merge [mɜːdʒ]	v.	to become one; mix together different elements 合并;使合并

Running the merge manager on a whole directory (tree) instead of a single file will make you aware of all files in that directory that have been added, deleted, or modified.

整个目录(树状结构),而不是单个文件中运行的合并管理器,会让你意识到,该目录下的所有文件,哪些是添加的,删除的,或者修改的。

reshape [ˌriːˈʃeɪp]	v.	to shape anew or differently 改造;再成形

If there is any good to come out of these crises, it is that we have an unprecedented opportunity to reshape the rules and institutions that guide the international economy.

如果说这次危机可以带来一些好处的话,那就是我们获得了一个空前的良机,去重塑引领国际经济体系的规则和机构。

vectorization [ˌvektəraɪˈzeɪʃən]	n.	向量化

This is only a simple example, demonstrating the basic idea behind auto-vectorization.

这只是一个简单的范例,演示了自动向量化背后的基本思想。

scalar ['skeɪlə(r)]	a.	having size but no direction 标量的;数量的

Business objects are the primary mechanism for representing business entities, enabling everything from a simple basic object with scalar properties to a large complex hierarchy or graph of objects.

业务对象是用于表示业务实体的主要机制,支持从具有标量属性的简单基本对象到大型复杂层次结构或对象图的任何实体。

kernel ['kɜːnl]　　　　　　　*n.*　the inner and usually edible part of a seed or grain or nut or fruit stone 内核

This command notes how long it takes for an application to complete and also measures how long it spends in user space and in kernel space.

这个命令会显示一个应用程序运行需要多少时间,并可以测量它在用户空间和内核空间各花费了多少时间。

stand-alone ['stænd əlaʊn]　*a.*　capable of operating independently(计算机)独立运行的;(公司、组织)独立的

The company is targeting this app at students and professionals, and marketing it as a replacement for stand-alone graphing calculators, which is clearly reflected in the price.

该软件目标用户为学生和职业人士,公司希望该软件可以代替独立的图形计算器,通过价格就可以看出这一点。

hybrid ['haɪbrɪd]　　　　　　*a.*　produced by crossbreeding 混合的

Hybrid cars can go almost 600 miles by refueling.

混合动力汽车加满油能行驶将近六百英里。

optimal ['ɒptɪməl]　　　　　*a.*　most desirable possible under a restriction expressed or implied 最佳的;最理想的

Now that winter has come and for many of us swimming is no longer an accessible form of cross-training, what would you suggest as the optimal cross-training exercise?

冬季已然来到,对我们多数人来说,游泳不再是一种可行的交叉训练方式,那么作为最佳的交叉训练方式你认为是什么呢?

saturate ['sætʃəreɪt]　　　　*v.*　to cause (a chemical compound, vapour, solution, magnetic material) to unite with the greatest possible amount of another substance 使饱和,使充满

This benefit becomes especially important when the number of people requiring medical care at the peak of the pandemic threatens to saturate or overwhelm health care capacity.

在流行病高峰期间需要得到医治的人数过多，令保健机构人满为患或者不堪重负之际，这个好处就显得尤为重要。

cluster ['klʌstə(r)] *n.* a grouping of a number of similar things 集群

The system deploys only one cluster with two servers in the New York area.

该系统仅在纽约地区部署一个集群和两个服务器。

decompose [ˌdiːkəm'pəʊz] *v.* to separate (substances) into constituent elements or parts 分解；使腐烂

We could decompose a complex system into collaborating parts to further simplify the modeling and development processes that could result in more reuse.

我们可以将复杂的系统分解为几个协作部分，进一步简化可以获得更多重用的建模和开发流程。

cache [kæʃ] *n.* a part of a computer's memory that stores copies of data that is often needed while a program is running 电脑高速缓冲存储器

If anything fails, then you might have left out some files in the cache manifest.

如果出现任何失败，那么可能是你在缓存清单中遗漏了一些文件。

affinity [ə'fɪnəti] *n.* the force attracting atoms to each other and binding them together in a molecule 亲和力；吸引力

And he says what sets him apart from the rest of the presidential field, his friend included, is the Haitian people's true affinity for him.

他补充道，海地人真实的亲和力会让自己脱颖而出，不仅在统辖的领域也包括在他的朋友们当中。

synchronization [ˌsɪŋkrənaɪ'zeɪʃn] *n.* the relation that exists when things occur at the same time 同步；同时性

Even though it is simple and it supports any physical device storage type, data migration/replication across systems and host instances synchronization across organization is not possible.

尽管它很简单而且支持任何物理设备存储类型，但是不可能实现跨系统数据迁移/复制和跨组织实例同步。

outperform [ˌaʊtpə'fɔːm] *v.* to be or do something to a greater degree 胜过；做得比……好

Academics pointed out that it was possible to outperform in the short term, simply by taking a lot of risk.

有学者指出,只要敢承担大风险,那么投资收益在短期内超过平均水平是可能的。

➤ *Phrases and Expressions*

come and go 来来往往,来来去去
have an impact on 对……有影响;对……造成冲击
as opposed to 与……截然相反;对照
as a consequence 因此,结果
be bound with 与……绑定
in all of... 在所有的……

➤ *Terminology*

at regional and global scales 在区域和全球范围内
European climate community 欧洲气候共同体

➤ *Proper Names*

NEMO (Nucleus of the European Model of the Ocean) 欧洲海洋模型
MPI (Message Passing Interface) 消息传递接口
OpenMP (Open Multi-Processing) 共享存储并行编程
OMP (Orthogonal Matching Pursuit) 正交匹配追踪算法
SIMD (Single Instruction Multiple Data) 单指令多数据结构
OPA 海洋动力学模拟软件
HPC (high-performance computing) 高性能计算
MFS (Mediterranean Forecast System) 地中海预报系统
CMCC (Euro-Mediterranean Center on Climate Change) 欧洲-地中海气候变化中心
HPC (Handheld Personal Computer) 掌上电脑
AMD (Advanced Micro Devices) 超微半导体公司

➤ *Translation*

1. The initial domain described along the latitude,... dividing the horizontal grid in a regular mesh. (Para. 2)
   沿着纬度、经度和深度坐标所描述的初始域扩展到在一个划分水平网格并行任务的规则网格中。
   (这里主要是指《国际高性能计算应用杂志》发表的研究成果。)

2. This parallelization approach is common... operations are performed at each point on the simulation grid. (Para. 3)

这种并行化方法在许多其他海洋和流体动力学模型中是常见的,因为它是一种利用在模拟网格上的每个点执行相同操作任务的自然方法。

3. This means that all of the triply... parallelized in the k (vertical/depth) dimension. (Para. 4)

这意味着所有的三重嵌套循环都将在k(垂直/深度)维度上并行。

4. This approach implies that during the initialization, ... domain decomposition strategy. (Para. 5)

这种方法意味着在初始化期间,每个进程必须根据预定义的域分解策略创建块。

5. For these reasons a version of... cache use and instruction-level parallelism. (Para. 7)

由于这些原因,使用重新排序的数组索引的欧洲海洋模型版本可能在缓存使用和指令级并行方面更能有效地利用硬件功能。

6. They can be used, as representative of... without changing the entire code. (Para. 8)

它们可以作为整个应用程序的代表,在不改变整个代码的情况下,了解创建的更新软件带来的好处。

7. The second domain is defined ... each architecture under test. (Para. 9)

为了使每个测试体系结构在计算节点上的可用内存能够饱和,第二个域名被定义出来。

8. On more than 120 threads performance drops off rapidly in this case. (Para. 11)

在这种情况下,超过120个线程的性能会迅速下降。

【这个句子中的 in this case 短语,在翻译时直接将其译为"在这种情况下"同时置于句首,采用对等的翻译方法。首先是因其结构简单,语义明了,再次因为其有对等的汉语表达。】

9. This idea is illustrated where the MPI sub-domain is shown decomposed into eight tiles. (Para. 12)

这个想法在消息传递接口子域被分解成八个块的位置得到印证。

10. Our experiments with individual, tracer-related kernels... the outer loop with OpenMP. (Para. 13)

我们对单独的、与跟踪相关的内核进行的实验表明,只需将外部循环与共享存储并行编程进行简单的并行操作即可获得最佳的绝对性能指标。

【这是个含有宾语从句的复杂句,从句中由包含丰富语义的介词短语构成。可采用将短小精悍、含义丰富的抽象词和词组翻译为较为具体的词或词组的具体译法进行翻译。这样,即降低了英汉语言间的差别,使译文更具体,也达到与原文表达一致的效果。】

 **Exercises**

1. Fill in the blanks with the proper given words, and then translate the sentences into Chinese.

performance	approach	demonstrate	pervasive
sub-domain	mechanism	parallelism	model
latitudinal	architecture	implement	suitability
loop	computation	vertical	blocking

1) The calculating domain is parted into several _____, and structured and unstructured grid are used separately in different sub-domain to solve the problem met in the process of grid building.

2) A design strategy for this type of migration includes the concepts of _____, synchronization, storage, elasticity, multi-tenancy, and security.

3) These same four phases are replicated in every type of delivery _____, whether it is a major release or a smaller maintenance service release.

4) It is also found that the ionosphere manifests the asymmetry in response to north- and southward horizontal winds and to day and night winds, and also shows _____ difference.

5) Most of us have at least one or two on a constant _____, repeating the same negative, and often untrue, drivel day in and day out.

6) The drawback of this _____ is that it only moves the rules themselves and registered columns, but not the metadata associated with them.

7) Authentication problems are probably the most _____ class of security problems if we ignore software bugs, meaning that choosing a reasonable authentication technology is important.

8) However, at the Analysis _____ level, decisions about where functionality will be implemented, in hardware, software, or workers, have not yet been made.

9) It would be very costly in terms of network bandwidth and _____ (for data serialization) to access all this data from the client for processing and then persist any changes.

10) The following examples should _____ why business rules and other requirements should be viewed separately.

11) So once you finish reading this letter, you may put some time aside to digest it and _____ it in your life.

12) The _____ axis shows us our projected return on investment. The horizontal axis shows our scorecard rating, from low to high.

13) Sick sinus can also be caused by scarring near the sinus node that's slowing, disrupting or _____ the travel of impulses.

14) The _____ of any technology for the demanded solution should be given much more weight while making the decision rather than the market trends.

15) In 2008, they built a "nest" of branches on the balcony of the Russian pavilion at the Biennale of _____ in Venice and have exhibited in Paris and Luxembourg.

16) As with the original form, however, there's a _____ issue if you have to perform nine lookups to retrieve each of the nine virtualized fields.

2. Translate the following passage into Chinese.

The Intel Xeon Phi struggled to match the performance of the traditional Xeon CPU as soon as the kernel involved the thread synchronization and load imbalance introduced by halo-exchange (MPI) operations. The results also demonstrate that, for the traditional architectures, the management of multiple threads within a node is less efficient than the management of Linux processes and the pure MPI implementation is good when the sub-domain is relatively big. The hybrid implementation outperforms the pure MPI version once the sub-domain size becomes small on both the Intel and IBM systems. However, on the IBM Power-based systems the pure MPI implementation was the most performant in all of our tests. Finally, considering all of the proposed approaches, we conclude that the introduction of OpenMP through the parallelization of the outer loop (over vertical levels) can be considered the best tradeoff between programmability, robustness and performance.

3. Translate the following passage into English.

与线程内存关联相关的另一种考虑是内存分配和初始化方式。在内核进程下,物理内存的分配发生在第一次写入它的地方(而不是在分配它的地方)。通常,在生成线程池之前,消息传递接口进程(主线程)分配并初始化数据结构。第一次接触时,数据被自然地分配到内存"附近"的主线程。然后,其他线程必须访问这些数据,这些数据被分配到离它们"很远"的地方,从而导致性能下降。更好的策略是在初始化阶段也生成线程,这样每个线程都可以初始化其数据,从而在"仿射"内存中自然地分配数据。

Unit 3
Marine Geographic Information

Geographic Information Systems

1. Geographic information systems (GIS) enable complex analysis of objects on the basis of their spatial locations. Similar objects are grouped into layers, and information about each object is maintained in a relational **database**. Objects are typically represented in **vector** format as points, lines, or **polygons**. **Aerial** images, such as satellite photos, also can be used in most GIS packages. There are GIS platforms that are meant for specific industries and purposes, including transportation planning, network analysis, and **logistics**. These transportation-specific extensions enable a user to compute shortest paths between origins and destinations that minimize time, length, or any other attribute (e. g. , accident rate, delays). Links may be selectively enabled and disabled, and time penalties may be assessed to represent delays associated with locks or other link types. The model outputs the number of trips **traversing** each link, as well as travel time between origins and destinations. Although these models are meant for highway planning, they may be adapted for use in other transportation modes.

2. This research makes an important first step in quantifying maritime risk of coastal sea **lanes** using GIS layers and databases maintained by the US Coast Guard (USCG), US Army Corps of Engineers (USACE), and the US Customs and Border Protection Agency. The databases mentioned previously serve as building blocks inside a transportation-specific GIS to calculate casualty rates along the US coast. The methodology developed is potentially useful to other research efforts, including understanding the impacts of proposed legislation (e. g. , repeal of the Jones Act), simulating coastal traffic in emissions studies, or planning for a long-term port disabling event (such as a labor dispute or natural disaster). Here, "casualty" refers to an **allision**, collision, or grounding, and "risk" refers to casualties per exposure metric (vessel trips).

3. The planned casualty rate methodology transformed from a link-based risk analysis to a cell-based grid analysis. Grids where trips were computed using data from the Waterborne Commerce of the United States were excluded from the results analysis to test the entrance and **clearance** data **methodology** of estimating trips and casualty counts. The observations in this section are all based on the entrance and clearance trip data. On closer examination, some trends appeared:

4. The numerous oil platforms in the Gulf of Mexico do not create significantly higher casualty rates in the Gulf of Mexico shipping channels. Although high-casualty-rate grids do exist along the Louisiana and Texas oilfields, the highest casualty rates are concentrated near the intersections of shipping lanes. Before performing the research, it was expected that a high number of vessel-platform allisions might be occurring in areas other than the designated shipping lanes (i. e. , vessels might be tempted to leave the shipping lanes to save time).

5. Most of the high-casualty-rate grid cells are located at entrances to major ports. The top 10 grids as far as casualty rates are the entrance to the port of San Francisco (outside the Golden Gate Bridge) , the Southwest Pass and approaches to the Port of New Orleans, Louisiana, San Diego, California, and Everett and Bellingham, Washington. Recreational vessels in these areas play a major role in the high number of casualties. Recreational vessels made up <2. 5% (overall) of allisions, collisions, and groundings from 2002 to 2009 (MISLE-reported events). However, >22% of reported casualties in Everett and Bellingham involved recreational vessels (MISLE data). Similarly, commercial fishing accounted for 10% of casualties in the entire database but almost 19% outside of Everett and Bellingham.

6. The entrance to the Galveston Bay (where the Intracoastal Waterway meets the Houston Ship Channel) had the cell with the greatest number of casualties (681) during the 28-year period. Silting conditions that require unique vessel maneuvers (e. g. , Texas chicken) and crossing situations with Intracoastal Waterway traffic likely contributed to this large number.

7. There are no cells in open water with high casualty frequencies. The highest number of casualties in a cell 75 to 100 mi from the coast is four (in the Gulf of Mexico oil fields). Coming closer to shore, no cell 50 to 75 mi from the coast has more than six casualties (also in the Gulf of Mexico oil fields).

8.   Unfortunately, it is not possible to directly compare casualty rates among ship types. This is because of the **misalignment** of vessel categories in the casualty and entrance and clearance data as well as the lack of detailed trip data on the inland waterways.

9.   Inland **towboats** and **barges** comprise an overwhelming majority of the casualty records (60% since 2002) in the case study database. However, this is to be expected, because towboats and barges maneuver through obstacles (locks and dams, bridges, other vessels) at close quarters throughout the duration of the trip. Deepwater ports also make use of sea pilots and docking masters to maneuver ships from the sea **buoy** to the terminal and the most **congested** part of the trip.

10.   To summarize the results of the cell-based analysis, it shows the top 10 port entrances in terms of casualty rate and casualty frequency (minimum 1,000 trips in each cell with at least 50 casualties from 1980 to 2007). Only the cell containing the Southwest Pass to the Lower Mississippi River appears in both columns. Marrero, Louisiana, is listed twice in the casualty frequency column because the area is located on the border of two high casualty-rate cells. Finally, casualties and vessel traffic were analyzed within the 100-mi buffer from the US coastline. It contains the number of casualties and trip miles based on the distance from shore. It is **intuitive** that casualties will occur more often closer to shore. However, the cumulative effect of coastal trips is very strong, with five times as much traffic (in terms of trip miles) between 25 and 50 mi from shore compared with less than 25 mi (including port arrivals). These trip miles are a rough approximation of shipping lanes and are not the actual paths taken by vessels. Future research should examine AIS data archives to find the preferred routes and exact coastal traffic levels.

11.   A methodology was presented for adapting a highway traffic assignment model for use in maritime transportation risk assessment. Specifically, existing vessel traffic volumes along US coasts were quantified using historical US Customs and Border Protection Agency entrance and clearance data. The traffic assignment model used in the research is capable of showing vessel traffic by any attribute contained in the entrances and clearances database (year, flag of registry, International Classification of Ships by Type, and draft). Such specific routing capabilities can serve as a valuable information resource for evaluating effects of expanded infrastructure (i. e., Panama Canal expansion), **dredging** ship channels to greater depths, and proposed legislation. The research is also transferable in the area of coastal and marine spatial planning. This refers to a planning process "for analyzing current and anticipated u-ses of ocean, coastal and Great Lakes areas". The process makes extensive use of GIS for identifying suitable areas for activities that minimize conflicts and **adverse** environmental impacts and balance competing demands for marine resources. The methodology and results (GIS layers) from this research could serve as an important **facet** to understanding the **waterborne** commerce patterns, interactions, and risks along US coasts.

12. Casualty rate was computed for all coastal and deep water port approach links. Be-cause of the geographic spread and long return period of casualties, a 100-mi mesh-based approach to quantifying casualty rates was adopted in favor of a link-based ap-proach. The calculation of casualties per million vessel trips is not a complete risk assessment picture. Additional data sets including weather data (severe weather, wind speed, and visibility) and consequences data (probability of a **spill** resulting from a vessel-platform allision), would make the risk assessment calculations more complete.

13. Additional data sets could be integrated with the routing results from the study. The Automated Mutual-Assistance Vessel Rescue System is a ship position reporting sys-tem used to coordinate search and rescue operations when a vessel broadcasts a **mayday** call. As the ships report in at least on departure, arrival, and once every 48 hours while at sea, the system's historical data would serve as a great source from which to **calibrate** the IWN. Similarly, large archives of AIS data are already being used in collision-prediction models. These data would serve to better **bench-mark** exact congestion levels within ports, near-miss situations, and high resolution port transit tracks. Archiving AIS data received in the Gulf of Mexico oilfields would be an interesting exercise to see how often shipping lanes are adhered to for vessel transits. Once more short-sea shipping services are offered and shipping schedules are established, the network could be used for performance tracking.

(1,438 words)

➤ *New Words*

database [ˈdeɪtəbeɪs]
n. an organized body of related information 数据库,资料库
Therefore, every symptom in the symptom database should have at least one or more solutions or reasons for the symptom.
因此,症状数据库中的每一个症状都应该至少有一个或多个对应于该症状的解决方案或起因。

vector [ˈvektə(r)]
n. a variable quantity that can be resolved into components 矢量
In so far as linear motion is concerned, a body is in equilibrium if the vector sum of the forces acting upon it is zero.
就线性运动而论,如果作用于严格物体上诸力的矢量之和是零,那么这个物体就处于平衡状态。

polygon [ˈpɒlɪɡən]
n. a closed plane figure bounded by straight sides 多边形;多角形物体
Each point of the polygon is added to the temporary shape on line 23, and if it's the first point in the polygon, its location is recorded as the placement for the text annotating that piece.
多边形的各个点都将被添加到第 23 行的临时图形中,并且如果它是多边形中的第一个点,则把其位置记录为注释该部分文本的位置。

aerial [ˈeriəl]
a. in or belonging to the air or operating (for or by means of aircraft or elevated cables) in the air 空中的;从飞机上的
Aerial photos always seem to be breathtaking and different, as they show all the objects on earth from a new perspective.
航拍照片看上去总是那么摄人心魄、与众不同,因为它们以崭新的视角呈现了地面上的物体。

logistics [ləˈdʒɪstɪks]
n. handling an operation that involves providing labor and materials be supplied as needed 后勤;后勤学
WHO continues to provide support in surveillance, water and sanitation, social mobilization and logistics together with the Ministry of Health.
世界卫生组织继续与卫生部一起在监测水和卫生设施、社会动员以及后勤方面提供了支持。

traverse [ˈtrævɜːs]
v. to travel across or pass over 穿过;来回移动
We traverse the desert by truck.

我们乘卡车横穿沙漠。

lane [leɪn]　　　　　 *n.* a narrow way or road 航道;航线

The northbound sea lane should remain open at most times.

大部分时间北行方向的海上航道应该保持通畅。

allision [əˈlɪʒən]　　　 *n.* the striking of one ship by another 一艘船被另一艘船撞之情形

There is no risk of allision at least for the next century or so (forecasting much beyond that is tricky), so we have no fear about it.

至少到下个世纪为止,我们无须担心有碰撞的危险(也没有预报的那么棘手)。

clearance [ˈklɪərəns]　 *n.* official permission to proceed(交通工具出入港口或出入境的)许可

Through this, UPS could get its packages cleared by the customs authorities even before their arrival into the airport, thereby saving a lot of time required for the customs clearance process.

通过这一系统,联合包裹甚至可以在包裹到达机场之前就向海关办理完通关手续,从而节省了大量办理海关通关手续的时间。

methodology [ˌmeθəˈdɒlədʒi] *n.* the branch of philosophy that analyzes the principles and procedures of inquiry in a particular discipline 方法学;方法论

No methodology in and of itself can fix these issues.

没有任何方法学可以解决这些问题。

misalignment [ˌmɪsəˈlaɪnmənt] *n.* the spatial property of things that are not properly aligned 不重合;未对准

The second awkward conclusion is that the highly subjective nature of assessing currency misalignment will make it very hard for America or the IMF to agree on whether a currency is out of line.

第二个令人尴尬的结论是,评估货币偏差中带有强烈的主观特点,这使得美国或是国际货币基金组织在一种货币是否出现偏差上很难达成一致。

towboat [ˈtəʊbəʊt]　　 *n.* a powerful small boat designed to pull or push larger ships 拖船

Permits could be issued to travel through locks at specified times, and those permits could be traded among towboat captains.

可以通过许可证的发放将船闸通行限定在指定时间内,

而且许可证也可以在拖船船长之间进行交易。

barge [bɑːdʒ]   *n.* a flatbottom boat for carrying heavy loads (especially on canals)驳船,平底载货船

As a result, Shanghai's business is a mix between serving its immediate hinterland and trans-shipping containers to travel by truck, rail or barge elsewhere in the Yangtze Delta.

因此,该集团的上海业务是一个混合体,既服务于其毗邻的内陆地区,又通过卡车、铁路或驳船将集装箱运至长江三角洲其他地区。

buoy [bɔɪ]   *n.* bright-colored; a float attached by rope to the seabed to mark channels in a harbor or underwater hazards 浮标,航标

Beside it another buoy looks like two cross-country skis with wire strung between them.

在它旁边,另一个浮标看上去像中间串着绳子的两块越野滑雪板。

congest [kənˈdʒest]   *v.* to become or cause to become obstructed 拥挤

With the rapid development of city economic and citifying course, the tendency of the increasing quantity of motor vehicle will lead to traffic flow uprising, traffic congest, and traffic jam.

随着城市经济的快速增长和城市化进程的加快,城市机动车数量呈快速增长的势头必然会造成道路交通流量猛增、交通拥挤和堵塞。

intuitive [ɪnˈtjuːɪtɪv]   *a.* spontaneously derived from or prompted by a natural tendency 直觉的;凭直觉获知的

He had an intuitive sense of what the reader wanted.

他能直觉地感到读者需要什么。

dredge [dredʒ]   *v.* to search (as the bottom of a body of water) for something valuable or lost(用挖泥船等)疏浚;疏浚,挖掘

A dredger is a boat or a device which is used to dredge rivers.

挖泥船是用来挖掘或捞取河底物质的船或设备。

adverse [ˈædvɜːs]   *a.* contrary to your interests or welfare 不利的

But the only requirement is that they notify the project if they suffer any adverse effects from their participation.

唯一的要求是,当他们因参与此项工程而受到任何不利影响时,一定要通知项目研究人员。

facet [ˈfæsɪt]   *n.* a smooth surface (as of a bone or cut gemstone)面;方面

Indeed, almost every facet of software development proffers at least one framework.

确实,几乎软件开发的每个方面都提供了至少一个框架。

waterborne ['wɔːtəbɔːn]   *a.* transported/supported by water 水运的;由水浮起的

It's no coincidence that the number of lost boats and lives is far higher for fishing than for any other type of waterborne industry.

商业捕鱼的沉船数量和渔民死亡率比其他任何水运行业都要高得多,其实这种局面的出现并非偶然。

spill [spɪl]   *n.* liquid that is spilled; the act of allowing a fluid to escape 溢出液,溅出;溢出量

President Obama says that crews have made progress in cleaning up the spill, but the job is not done.

奥巴马总统说,相关人员在清理泄漏中已经取得进展,但这项工作尚未结束。

mayday ['meɪdeɪ]   *n.* an internationally recognized distress signal via radiotelephone (from the French m'aider) 求救信号

But even if another vessel is involved, it's hard to explain how an alert captain or watch stander would not have spotted an approaching container ship in time to make an adequate mayday call.

即便真的有其他船只牵涉进来,也很难解释为什么一位警觉的船长或值班人员都没有及时发现这样一艘正在靠近的集装箱货船并发出求救信号。

calibrate ['kælɪbreɪt]   *v.* to make fine adjustments or divide into marked intervals for optimal measuring 校正;校准

From the Moon, Earth is an ideal reference point to calibrate the cameras.

从月球的角度看,地球是一个理想的校准相机的参照点。

benchmark ['bentʃmɑːk]   *n.* a standard by which something can be measured or judged 基准;标准检查程序

Unlike America, Europe seems keen to use a lax definition of capital that is no longer viewed by the market as the best benchmark for solvency.

与美国不同,欧洲似乎热衷于使用对于资本的一个模糊定义。然而市场已经不认为这一定义是衡量偿债能力的最好基准。

➤ *Phrases and Expressions*

on the basis of 根据;基于……
enable... to do... 使……要做……
as well as 也;和……一样;不但……而且……
account for 对……负有责任;对……做出解释;说明……的原因
serve as 担任……,充当……;起……的作用

➤ *Terminology*

between origins and destinations 起点与终点之间
open water 开阔水面;无冰水面

➤ *Proper Names*

GIS（Geographic Information Systems）地理信息系统
USCG（US Coast Guard）美国海岸警卫队
USACE（US Army Corps of Engineers）美国陆军工程兵团
US Customs and Border Protection Agency 美国海关和边境保护局
the Gulf of Mexico 墨西哥湾
the Galveston Bay 加尔维斯顿湾
AIS（Automatic Identification System）船舶自动识别系统
IWN（international Waterway Network）国际航道网络
the Golden Gate Bridge 金门大桥(美国旧金山市的著名建筑物)

➤ *Translation*

1. Similar objects are grouped into layers, and... each object is maintained in a relational database. (Para. 1)
   类似的对象被分组到层中,并在相关数据库中维护关于每个对象的信息。
   (这里主要是指《交通研究记录》杂志的海洋调查研究。)

2. Links may be selectively enabled and disabled, and time... to represent delays associated with locks or other link types. (Para. 1)
   可以有选择地启用和禁用链接,并评估时间损失,来表明与锁或其他链接类型的延迟。

3. Here, "casualty" refers to an allision, collision... refers to casualties per exposure metric (vessel trips). (Para. 2)
   这里,"伤亡"指的是相撞、碰撞或搁浅,而"风险"指的是每曝光量度的伤亡数(船舶航行期间)。
   【这个句子采用直译方法进行翻译,保持原文内容又保持了原文形式。前半句中,宾语是三

个并列名词,依次翻译,后半句的"and"一词翻译为转折连词"而","vessel trips"译为"船舶航行期间"增加了词汇"期间"的增词译法,使句子通顺自然,便于理解。】

4. This is because of the misalignment of vessel categories... the lack of detailed trip data on the inland waterways. (Para. 8)

这是由于船舶类别在伤亡和出入港数据上的不一致,并且内河航道缺乏详细的航行数据。

5. Deepwater ports also make use of sea pilots... the most congested part of the trip. (Para. 9)

深水港口还利用海上引航员和码头管理员来操纵船只从海上航标到码头,以及航行中最拥挤的区域。

6. Marrero, Louisiana, is listed twice in the casualty... two high casualty-rate cells. (Para. 10)

路易斯安那州的马雷罗在伤亡率一栏中被列出两次,因为该地区位于两个高伤亡率位置的交界处。

7. These trip miles are a rough approximation... actual paths taken by vessels. (Para. 10)

这些航行里程只是粗略近似于海运航线,并不是船只实际行驶的路线。

8. A methodology was presented for adapting... for use in maritime transportation risk assessment. (Para. 11)

提出了一种将公路交通分配模型应用于海上运输风险评估的方法。

9. The process makes extensive use of GIS... impacts and balance competing demands for marine resources. (Para. 11)

这一过程广泛利用地理信息系统来确定适当的活动区域,尽量减少冲突和不利的环境影响,并平衡对海洋资源的竞争性需求。

10. Casualty rate was computed for all coastal and deep water port approach links. (Para. 12)

计算所有沿海和深水港进港航道的伤亡率。

【这个句子中的被动语态"was computed"译为主动语态更为合适,易于读者接受,这是转译法的翻译方法。句末的"deep water port approach links"译为"深水港进港航道"更符合句意,使用了具体译法的翻译方法,降低了英汉间的语言差别。】

Exercises

1. Fill in the blanks with the proper given words, and then translate the sentences into Chinese.

frequency	congest	maneuver	complex
spatial	casualty	transportation	link
clearance	lane	recreational	channel
towboat	buffer	assessment	network

1) Every location, of course, is its own neighbor by this query's _____ search criterion, so exclude that row from the search.

2) This year, the school added a new master's program in _____ design, one of only a few in the country, that will combine business classes with design.

3) It is vital to solve the traffic _____ problems that researching deeply on dynamics characteristic of traffic flow in our country and establishing new-style traffic flow theory.

4) You can write trace data to a file continuously as an extension to the in-storage trace, but instead of one buffer per thread, at least two _____ per thread are allocated.

5) My next step is to wire them up using a common _____ as a shared synchronization mechanism and then kick each one off.

6) Rational, unemotional, self _____ should tell us whether or not we have the skills and ability to do something.

7) Where everyone else was fixated on trying to placate international investors, Iceland imposed temporary controls on the movement of capital to give itself room to _____.

8) Because of the decoupled approach we discussed above, this does not have an impact on the _____ with which the service request against the target can be executed.

9) The collective intelligence of an ant colony can serve as inspiration to help us solve _____ human problems.

10) Each request carries some data over the _____; if the request does not find any updates on the server, this data transfer is a waste.

11) With the development of _____ and barges large enough to hold almost an entire trainload of cargo, however, the port grew to be second in the nation after World War Ⅱ.

12) The most common suite of driver assistance technologies available today includes adaptive cruise control, _____ departure warning systems, and parking assistance systems.

13) Russia with more forest than any other country also lost a lot, and the FAO's figures do not capture because its _____ did not involve a permanent change in land use.

14) This *American Express Spending & Saving Tracker* surveyed consumers about their plans and spending intentions as they relate to summer travel and _____ activities.

15) At that hour, midday Friday, the US Navy hospital ship *Comfort* had temporarily suspended flights for incoming patients because the ship's _____ receiving department was beyond capacity.

16) If they understand how you make money, they can understand the _____ between their performance and the bottom line.

2. Translate the following passage into Chinese.

Additional data sets could be integrated with the routing results from the study. The Automated Mutual-Assistance Vessel Rescue System is a ship position reporting system used to coordinate search and rescue operations when a vessel broadcasts a mayday call. As the ships report in at least on departure, arrival, and once every 48 h while at sea, the system's historical data would serve as a great source from which to calibrate the IWN. Similarly, large archives of AIS data are already being used in collision-prediction models. These data would serve to better benchmark exact congestion levels within ports, near-miss situations, and high resolution port transit tracks. Archiving AIS data received in the Gulf of Mexico oilfields would be an interesting exercise to see how often shipping lanes are adhered to for vessel transits. Once more short-sea shipping services are offered and shipping schedules are established, the network could be used for performance tracking.

3. Translate the following passage into English.

在全球范围内,发展包含技术的通信网络充满了权力、政治归属、冲突、外交成败及经济考量等问题。殖民时期的英国参与了所有上述过程,同时建立了电报和无线网络系统,巩固了她在当时成为强大国家的地位。这种巩固并非仅仅来自确定性的调控机构,而且还得到了当地利益攸关者的帮助,他们将从为其自身利益和议程而建立起来的新通信技术中获得收益。贝利曾指出,"在1857年,尽管更多的道路和邮政系统比电报扮演了重要的角色,但扮演合作角色的精英们在殖民地建立和使用了新的通信网络技术"。

Chapter 8

Ocean Economy

Unit 1

Blue Economy

The Blue Economy:
Growth, Opportunity and a Sustainable Ocean Economy

1. Our economic relationship with the ocean is evolving in important ways. As a setting for global trade and commerce, and as a significant source of food and energy, the ocean's contribution is important.

2. The term "industrialization" is sometimes used to signal this gathering trend of expansion and **acceleration** of human activity in and around the ocean. It is probably not far from the truth. Alongside established ocean industries, emerging and new activities-offshore renewable energy, **aquaculture**, deep seabed mining and marine **biotechnology** are often cited-will bring new opportunities, growth and greater diversity to the ocean economy. Governments too are playing a key role in driving growth. Through new national ocean development plans, countries are turning to the ocean as a "new" source of jobs, innovation and competitive advantage.

3. There is another, very important dimension to future growth in the ocean, and that is the so-called "blue economy" (or "blue growth"). The concept has its origins in the broader green movement and in a growing awareness of the heavy damage **wrought** on ocean ecosystems by human activity such as overfishing, habitat destruction, pollution and the impact of climate change. A **tinge** of "blue" can be found in most new national ocean strategies and policies as governments signal an intention to promote a more sustainable balance between economic growth and ocean health. Much of the wider ocean discourse has also gone "blue"—even if there are plenty of different views of the concept and there is no widely agreed or consistent definition.

4. The world is slowly waking up to this—and it may not be too late, either. Some argue, quite reasonably, that the extent of the damage to the ocean is many decades shy of the impact of industrialization on land, and there is still time, if we act now, to get the principles and the framework for the development of the ocean economy right.

5.   But can the concept of the blue economy be more than **aspirational**? What do governments, and others, really mean when they refer to it? Is the blue economy at risk from
"greenwashing"? What are the challenges of "blueing" the ocean economy?

6.   The ocean is becoming a new **focal** point in the discourse on growth and sustainable
development, both at national and international levels. The world is in many ways at a
turning point in setting its economic priorities in the ocean. How this is done in the
next years and decades, in a period when human activities in the ocean are expected
to accelerate significantly, will be a key determinant of the ocean's health and of the
long-term benefits derived by all from healthy ocean ecosystems.

7.   The idea of the "blue economy" or "blue growth" has become synonymous with the
"greening" of the ocean economy. Increasingly, national ocean development strategies reference the blue economy as a guiding principle, but it is not always clear what
a sustainable ocean economy should look like, and under what conditions it is most
likely to develop.

8.   A wide array of thinking is being brought to bear on the blue economy, and it is a sure
sign of progress that many countries are now adopting or exploring sustainable frameworks for their ocean economies. But the difficult work has only just begun. Defining
the blue economy, then **articulating** how to piece together the enabling legal, governance, investment and financing arrangements-and implementing these-will be a major
challenge.

9.   The ocean will become an economic force this century.

10.   The economic contribution of the ocean is significant but remains undervalued.
Measuring the ocean economy gives a country a first-order understanding of the economic importance of the seas. Measurement is difficult, not least because the lines
between coastal and ocean economies are often **blurred**. And comparisons between
countries are complicated by differences in measurement systems, income and geography. It is also likely that the economic contribution of the ocean is undervalued in
many countries. National accounting systems often treat large sectors, such as oil
and gas and coastal real estate separately. Meanwhile, few estimates give any sense
of the value of non-market goods and services, such as carbon **sequestration**, to
the ocean economy.

11.   A new wave of "industrialization" of the ocean and coasts is under way, the scale of
which is only now becoming apparent. Trends point to accelerating economic activity in and around the ocean, against the **backdrop** of a **soaring** global population,
growing **affluence** and consumption, and the need for new sources of food, energy

and minerals. On land, the ocean-related economy will experience a **surge** in investment in coastal infrastructure, industry and tourism as the global migration to cities and coasts, deepens. At the same time, the risks to coastal populations of rising sea levels and storm surges as a result of climate change will drive a wave of defensive infrastructure development.

12. A strategic focus on the development of national ocean resources will be an important driver and enabler of the ocean economy. Ocean economies, both large and small, are looking to their seas to **bolster** slowing growth in their **terrestrial** economies, discover new opportunities for investment and employment, and build competitive advantage in emerging industries such as deep seabed mining and marine biotechnology. New strategic ocean development plans and policies, sometimes referred to as "blue economy" plans, are being drafted to stimulate growth in and around countries' exclusive economic zones (EEZs). Should these public policy ambitions prove successful at enabling investment, the scale, size and type of economic activity in the ocean will be of an entirely new order.

13. Human activities are driving the decline in ocean health.

14. There has been a shocking **plunge** in ocean health that has been directly linked to human activities. The "great acceleration", the extraordinary burst of economic and industrial activity that began in the second half of the 20th century, has led to **unprecedented** changes in ocean ecosystems. The urgency of the ocean health challenge is becoming more prominent in the global policy discourse. For the first time the ocean is on the G7 agenda in 2015. In the two decades since the first Earth Summit in 1992, the efforts of the conservation and science communities, as well as a handful of visionary political and business leaders, have succeeded in elevating the issue of ocean health to the highest levels of global policymaking. The opportunity for meaningful progress has never been greater, even if the record suggests cautionary optimism.

15. The blue economy is as yet a new **paradigm** in name only.

16. The emerging concepts of the blue economy and blue growth are important public policy aspirations. With sustainable growth the new focus of the global policy discourse, countries seeking to develop their ocean economies have, to varying degrees, acknowledged the need for policies that better align future economic growth in their seas with maintaining or even restoring ocean health. The terms blue economy and blue growth, used liberally in national ocean plans, imply just this-a measure of greening of the ocean economy. Welcome as this development is, these emerging concepts—and **counterparts** such as "sustainable ocean economy"—remain

ill-defined and open to wide, and often different, interpretations. Fears of "blue-washing" **abound**.

17.　The blue economy typically prioritizes growth over sustainability. The idea that the blue economy is, in the minds of most policymakers, a relatively conventional "ocean economy" seems to be borne out by a more careful national ocean development plan. While the concept of the blue economy links economic growth with the conservation of ocean ecosystems, it seems clear that neither the conservation or sustainability component is the primary nor even necessarily the ultimate goal.

18.　Without a common understanding of the blue economy and a clear framework for sustainable growth, even modest progress on ocean health will be a challenge. While governments are clearly receptive to improving ocean health and building a sustainable ocean economy, the tendency to prioritize growth, and the prevailing view that a few judicious changes to policy, governance and enforcement should be sufficient to manage the impact of growing competition for ocean space, suggest that the scale of the ocean challenge is not yet fully appreciated. Bringing greater clarity to the concept of the blue economy will be essential, but equally important is a robust framework for governance, institutions, investment and business innovation that will **underpin** a "bluer" economy.

19.　The transition from an ocean economy to a blue economy will be a complex, long-term undertaking.

20.　A sustainable ocean economy offers a path for considering economic development and ocean health as **compatible propositions**. It does not have to be a choice between growth and sustainability. Properly planned and managed ocean spaces should mobilize public and private sector investment and generate strong returns and ecosystem benefits. The advantages of such an approach mean that a diversity of activities, from traditional ocean sectors to new businesses focused on ocean health, can be managed in a coordinated way, within a comprehensive framework of ecosystem-based management.

21.　A clear policy is essential, but not sufficient. The devil, as always, is in the detail. Reform of institutions governing the ocean economy is required to keep pace with accelerating economic activity. Ecosystem-based management in which both the economy and ecosystems thrive, and its most important implementing tool, marine spatial planning (MSP), require a set of integrated governance and supporting conditions to be present.

(1,558 words)

➤ *New Words*

sustainable [səˈsteɪnəbl]　*a.* that can continue or be continued for a long time 可持续的；可以忍受的；足可支撑的；养得起的

Sustainable development is all about creating better health care, education, housing and improved standard of living for everyone.

可持续发展就是要为每个人创造更好的医疗、教育、住房条件，提高人们的生活水平。

acceleration [əkˌseləˈreɪʃn]　*n.* an increase in rate of change 加速，促进

They wanted stronger protection for their patents against unfair competition, and acceleration of the product-approval process.

他们希望自己的专利能得到更加强有力的保护，来应对不公平竞争，还希望加快产品认证进程。

aquaculture [ˈækwəkʌltʃə(r)]　*n.* rearing aquatic animals or cultivating aquatic plants for food 水产养殖；水产业

The powerhouse of aquaculture in the world is Asia—Southeast Asia and China in particular.

全球水产养殖最发达的地区就是亚洲，特别是南亚和中国。

biotechnology [ˌbaɪəʊtekˈnɒlədʒɪ]　*n.* the branch of molecular biology that studies the use of microorganisms to perform specific industrial processes 生物技术；生物工艺学

Biotechnology provides powerful tools for the sustainable development of agriculture, fisheries and forestry, as well as the food industry.

生物技术为可持续农业、渔业和林业发展以及食品加工业提供了强有力的手段。

wrought [rɔːt]　*v.* to have caused something to happen, especially a change 使发生了，造成了（尤指变化）

Irresponsible industrial practices have wrought terrible environmental abuses in forms of air, land and water pollution.

不负责任的工业行为导致了空气、土地和水污染，对环境造成了严重破坏。

tinge [tɪndʒ]　*n.* a small amount of a color, feeling or quality 微量，少许，一丝，几分（颜色、感情或性质）

We instinctively feel a tinge of pain when we observe another in pain—at least most of us do.

看到他人痛苦,我们就会本能地感到一丝疼痛,至少大部分人都是如此。

aspirational [ˌæspəˈreɪʃənl]　　*a.* aimed at or appealing to people who want to attain a higher social position or standard of living(社会地位或者生活水平等方面)梦寐以求的;有雄心壮志的

The goals of near-zero deaths and eventual eradication are idealistic and aspirational.

将死亡数降至接近零并最终将其消灭,这是梦寐以求的理想目标。

focal [ˈfəʊkl]　　*a.* having or localized centrally at a focus 焦点的;有焦点的;中心的;很重要的

He quickly became the focal point for those who disagreed with government policy.

他迅速成为异见人士的中心人物。

articulate [ɑːˈtɪkjuleɪt]　　*v.* to speak, pronounce, or utter in a certain way 明确表达;清楚说明

We have postgraduates at interview who can't articulate why they decided to do the course, nor what added-value they gained from it.

我们访谈时遇到过一些研究生,他们既说不清楚为什么会选这门课程,也说不出从中得到了什么。

blur [blɜː(r)]　　*v.* to make less distinct or clear(使)变得模糊不清;(使)视线模糊;(使)看不清;(使)难以区分

Besides, smaller utility vehicles are coming along that truly blur the differences among cars, vans and trucks.

此外,出现了越来越多的小型越野车,确实模糊了轿车、货车和卡车之间的区别。

sequestration [ˌsiːkwəˈstreɪʃn]　　*n.* the process of capture and long-term storage of something 封存;扣押;隔离;没收

Obesity is the body's way of storing fats and that metabolic sequestration actually protects other organs from the harmful effects of fat.

长胖是人体储存脂肪的一种途径,这种代谢贮藏可以有效防止其他器官受到脂肪的不良影响。

backdrop [ˈbækdrɒp]　　*n.* the general conditions in which an event takes place 背景;背景幕

Regardless of the particular backdrop or specific situation, most people have an underlying belief in fairness.

无论背景如何,情况如何,大多数人对于公平还是有着基本信念的。

soar [sɔː(r)]   *v.* to rise rapidly 急升;猛增;升空

You soar a mile above the ocean, then slowly circle for 20 minutes as the pilot guides you to a leisurely landing on the beach.

你急升到海面 1 英里高空处,然后慢慢盘旋 20 分钟,最后在飞行员的指导下从容地降落在海滩。

affluence [ˈæfluəns]   *n.* the state of having a lot of money or a high standard of living 富裕;丰富

The upsurge of divorce in the 1970s was caused by new social mores and legislative changes rather than affluence.

20 世纪 70 年代,离婚率突然飙升,这是因为社会习俗和立法都发生了变化,而不是因为大家有钱了。

surge [sɜːdʒ]   *n.* a sudden large increase in something (数量的)急剧上升,激增

With the surge in digital sales, publishers are casting a worried eye towards the self-published market.

随着数字图书销量的激增,出版商们开始对自助出版市场感到担忧。

bolster [ˈbəʊlstə(r)]   *v.* to support and strengthen 改善;加强

To bolster its overall competitiveness, Samsung has been focusing on improving its smartphone offerings.

为了增强整体竞争力,三星一直专注于改善其智能手机产品。

terrestrial [təˈrestrɪəl]   *a.* operating or living or growing on land 陆地的;地球的

Aquatic ecosystems share a number of basic structural and functional characteristics with terrestrial ecosystems.

在结构和功能方面,水生生态系统和陆地生态系统有很多相似特征。

plunge [plʌndʒ]   *n.* a sudden decrease in an amount or the value of something 突然跌落;(价格、数量等)骤降;投入

He argues that a plunge on Wall Street might scare investors away.

他认为,华尔街股市暴跌可能会吓跑投资者。

unprecedented [ʌnˈpresɪdentɪd]   *a.* that has never happened, been done or been known before 空前的;无前例的

At the end of World War Ⅱ, the US faced unprecedented international challenges.

第二次世界大战结束时,美国面临着前所未有的国际挑战。

paradigm [ˈpærədaɪm]   *n.* a typical example or pattern of sth 模式;典范;范例

The economic paradigm is in shambles that predominated in the years before the crisis.

危机发生的前几年里,这种经济范式曾经是主流。

counterpart [ˈkaʊntəpɑːt]   *n.* a person or thing having the same function or characteristics as another 对应的事物;职位(或作用)相当的人

Europe's computer industry is usually playing catch up with its faster-moving US counterpart.

与飞速发展的美国同行相比,欧洲的计算机行业一直都是个跟班的角色。

abound [əˈbaʊnd]   *v.* to be abundant or plentiful; exist in large quantities 大量存在,有许多;富于,充满

Worrywarts abound in international finance circles—there's always nail-biting about this currency fluctuation or that account deficit.

国际金融界从来都不乏杞人忧天之人,他们一天到晚坐立不安,不是担心汇率波动,就是担心账户赤字。

underpin [ˌʌndəˈpɪn]   *v.* to support from beneath 巩固;支持;从下面支撑;加强……的基础

This engagement will underpin our commitment to a new international order based upon rights and responsibilities.

这项工作将有助于我们以权利、义务为基础建立国际新秩序。

compatible [kəmˈpætəbl]   *a.* able to exist and perform in harmonious or agreeable combination 兼容的;能共处的;可并立的

His task was to elaborate policies that would make a market economy compatible with a clean environment.

他的任务是制定完善政策,既保障市场经济发展,又保证不污染环境。

proposition [ˌprɒpəˈzɪʃn]   *n.* the act of making a proposal 命题;提议;主题;议题

Any suggestion or proposition helpful to the peaceful solution of the nuclear issue on the Peninsula should be discussed carefully and fully by all the parties.

所有议题,只要是有利于和平解决朝鲜半岛核问题的,相关各方都应给予充分、认真的讨论。

➤ *Phrases and Expressions*

wake up to 认识到,意识到

be shy of 畏缩;羞于……

at risk 处于危险中

derive... from 源出,来自,得自;衍生于

an array of 一排;一批;大量

give a sense of 给一种……感觉

be under way 进行中;正在开展

align with 与……结盟

be borne out 被证实

keep pace with 保持同步,与……齐步前进

➤ *Terminology*

greenwashing 合成新词,green(象征环保)和 whitewash(漂白),意指某些公司、政府或是组织为树立环保形象而进行虚假宣传,而实际却反其道而行。

real estate 房地产

carbon sequestration 固碳技术,也称碳封存,捕获碳并以安全的方式进行存储,而不是直接向大气中排放,包括物理固碳和生物固碳:物理固碳是将二氧化碳长期储存在开采过的油气井、煤层和深海里,生物固碳是将二氧化碳转化为有机碳即碳水化合物,固定在植物体内或土壤中。

storm surge 风暴涌浪,热带气旋不断靠近大陆的过程中造成海平面上升形成浪涌(注意与 strom tide 风暴潮区分)

blue-washing 漂蓝,源自 greenwashing

➤ *Proper Names*

Blue Economy 蓝色经济

EEZ (Exclusive Economic Zone) 专属经济区

G7 (Group of Seven) 七国集团

Earth Summit 地球高峰会议,又称"联合国环境与发展会议"(简称 UNCED)、里约热内卢高峰会,联合国重要会议之一

MSP (Marine Spatial Planning) 海洋空间规划

➤ *Translation*

1. Our economic relationship with the ocean is evolving in important ways.（Para. 1）
   人类与海洋之间的经济关系正在经历非常重大的变革。

2. As a setting for global trade and commerce, and as a significant...（Para. 1）
   海洋是全球贸易和商务活动的主要场所,也是食物和能源的重要来源,它对人类的贡献意义非凡。

3. Increasingly, national ocean development strategies reference...（Para. 7）
   各国的海洋发展战略越来越多地将蓝色经济作为指导思想,但是,对于可持续海洋经济到底应该是什么样的,发展可持续海洋经济到底需要什么条件,却没有一个清晰的认识。

4. And comparisons between countries are complicated by differences...（Para. 10）
   要将所有国家放在一起比较本来就不是一件容易的事,加上各国的评估体系不一样,国民收入和地理条件也各不相同,就更困难了。
   【首先,将"comparison"转译为动词,同时将其从主句中拆分出来单独成句,以符合汉语板块式结构的特点要求;其次,根据上下文以及"are complicated by"这个表达所体现出来的语义增译"本来就不是一件容易的事",同时将作为动词使用的"complicate"转译为符合汉语表达习惯的形容词;另外,将名词"difference"转译为形容词,并根据不同搭配翻译为"不一样"、"各不相同"以避免重复。】

5. Meanwhile, few estimates give any sense of the value of non-market...（Para. 10）
   而且,在对海洋经济进行评估的时候,很少能考虑到类似固碳技术这种非卖产品或服务的价值。

6. At the same time, the risks to coastal populations of rising sea levels...（Para. 11）
   同时,由于气候变化,导致海平面上升、风暴涌浪灾害增加,给沿海居民的生活带来了威胁,这也会掀起一波防御性基础设施的建设高潮。

7. Should these public policy ambitions prove successful at enabling...（Para. 12）
   如果推出的这些政策能够达到其目的,成功地吸引到投资,那么海洋经济活动无论从规模、还是类型,都将呈现出全新的面貌。

8. The "great acceleration", the extraordinary burst of economic and...（Para. 14）
   自 20 世纪下半叶开始,出现了"大提速"现象,经济和工业活动呈现出不同寻常的繁荣景象,这也导致了海洋生态系统发生了前所未有的变化。

9. The opportunity for meaningful progress has never been greater...（Para. 14）
   尽管从现有记录看来,目前还只能保持谨慎乐观,但未来还是有可能取得重要进展的,而且这种可能性比以往任何时候都大。

10. Welcome as this development is, these emerging concepts and...（Para. 16）
    对于这种发展趋势,大家都表示支持,尽管如此,对于这些新概念,还有"可持续海洋经济"

等这些类似的新名词,人们的定义往往不太妥当,而且对其理解和诠释也各不相同,颇有争议。

【英语书面语中常用物称主语,也称无灵主语,意在使叙述显得客观、公正,语气较为委婉、间接,而汉语则较注重主体思维,往往从人出发来叙述客观事物,因而常用人称主语,因此,在翻译中放弃"development"这个物称主语,增译虚指主语"大家",将"defined"转译为名词,同时增译虚指主语"人们",以符合汉语表达习惯。】

11. Without a common understanding of the blue economy and a clear... (Para. 18)

如果没有就蓝色经济的概念达成共识,也没有制订一个清晰的可持续发展框架,那么想要提升海洋生态环境质量,即便是小小的提升,也是非常困难的。

【翻译该句主要使用转译和拆分这两个重要技巧:首先,在"understanding"和"framework"前分别增译"达成""制订",从而将这两个名词结构扩充成为译文中的两个分句;其次,将"progress"转译为动词"提升",扩充成为第三个分句;另外,将"even modest progress"这个结构单独译为"即便是小小的提升",以突出"even"在英语中作为强调表达的作用;最后,将"challenge"转译成为形容词,以符合汉语表达习惯。】

12. Ecosystem-based management in which both the economy and... (Para. 21)

以保护生态环境为前提进行经济活动管理,既保证经济繁荣,又保证生态健康,一个重要的实施手段就是进行海洋空间规划,这些都需要出台一整套管理办法和一系列支持措施。

 **Exercises**

1. *Fill in the blanks with the proper given words, and then translate the sentences into Chinese.*

sustainable	diversity	promote	wake up to
at risk	derive... from	give a sense of	be under way
ambition	unprecedented	conservation	acknowledge
interpretation	prevailing	appreciate	keep pace with

1) The museum contains numerous utensils, rock paintings and carvings which _____ the archaic lives of these peoples.

2) Differences between languages _____ the different modes of thinking as thinking is the basis of language.

3) One often-posed question about multinational corporations relates to the nature of the relationship between the cultures of corporations and the _____ national cultures.

4) A key issue for higher education in the 1990's is the need for greater _____ of courses provided for the students.

5) The Chancellor emphasized his determination to _____ openness and transparency in the

government's economic decision-making.

6) He will inevitably put his administration _____ if he doesn't come through on these promises for reform.

7) More important, international investors _____ the relative weakening of America's economic power.

8) According to the World Bank, between 2005 and 2055 agricultural productivity will have to increase by two-thirds to _____ rising population and changing diets.

9) The developing countries also need to take due adaptation actions in the context of _____ development.

10) If you know something about the author's family background, you will _____ the poem on an even profounder level.

11) According to the report, even if the recovery _____ , it may be some time before the official statistics confirm it.

12) As a well-known proverb goes: "There are a thousand Hamlets in a thousand people's eyes", a good novel should be open to _____ .

13) As time went on, his _____ to be part of the US Supreme Court faded in alcohol and despair.

14) _____ is an issue which gets a lot of attention these days—whether it means preserving old buildings, or protecting the environment.

15) China's education and training industry has undergone a(an) _____ development with its industrialization and international development process.

16) I now realize that I should have had the courtesy to _____ my rudeness and apologize, but words failed me at the time.

2. Translate the following passage into Chinese.

There is another, very important dimension to future growth in the ocean, and that is the so-called "blue economy" (or "blue growth"). The concept has its origins in the broader green movement and in a growing awareness of the heavy damage wrought on ocean ecosystems by human activity such as overfishing, habitat destruction, pollution and the impact of climate change. A tinge of "blue" can be found in most new national ocean strategies and policies as governments signal an intention to promote a more sustainable balance between economic growth and ocean health. Much of the wider ocean discourse has also gone "blue"—even if there are plenty of different views of the concept and there is no widely agreed or consistent definition.

3. Translate the following passage into English.

发展蓝色经济指的是发展依赖于海洋或与海洋相关的环境友好型产业和经济。发展蓝色经济对中国这样的海洋大国至关重要,今后蓝色经济在国民经济中所占比重会越来越高。此次联合国海洋大会讨论了海洋污染、海洋生态保护、海水酸化、可持续渔业、海洋科研能力等议题,中国向大会提交了5份自愿承诺,在加强海洋生态的管理和保护、提高防灾减灾能力、加强科技创新、发展可持续的海洋经济以及推动亚太区域海洋国际合作5个方面提出了发展目标。

Unit 2
Issues and Challenges

Issues on the Blue Economy

1. The natural world made up of the physical environment, its mineral components and biodiversity at all three levels (genetic, species, ecosystem) is intrinsically interconnected and the more diverse and productive the natural system, the greater the degree of interconnectivity. Hence the identification of particular issues is inherently an **anthropogenic construct** and depending on one's perspective may appear arbitrary. A case in point is the **precursory** role that the conservation and sustainable use of biodiversity has in enabling the establishment of a Blue Economy, broader sustainable development and poverty eradication. This is particularly true in developing countries where economies are more directly related to environmental exploitation.

2. Equivalent figures could be developed for the other issues, underlining the overall interconnectivity and the need for an integrated and holistic approach. To this end the ecosystem approach must underpin all aspects of the Blue Economy incorporating inter-relationships, **knock-on** effects, externalities and the true costs and benefits of activities in terms of the natural blue capital.

3. Sustainable use of biodiversity.

4. The natural capital of many marine and coastal ecosystems has been degraded, impacting upon the provision of services and livelihoods. Approximately 20% of the world's coral reefs have been lost and another 20% degraded. **Mangroves** have been reduced to 30%–50% of their historical cover and it is estimated that 29% of seagrass habitats have disappeared since the late eighteen hundreds.

5. An ecosystem approach is required that factors in restoration of biodiversity and renewable resources, and proper management of resource extraction. For example, in fisheries, some of the renowned "Sunken Billions" could be restored providing the basis for productive, efficient, sustainable fisheries and enhanced food security. The sci-

entific determination and designation of appropriate MPAs can play a key role in this regard reconstituting biodiversity, ecosystem services and general resilience to other system shocks. Currently only some 2% of our oceans are protected, despite the CBD/WSSD 2012 target of a representative 10% area, whereas approximately 12% of terrestrial areas are under protection.

6. Food security.

7. In the context of the Blue Economy food security is very closely related to the sustainable use of biodiversity particularly where it **pertains** to the exploitation of wild fisheries. 1 billion people in developing countries depend on seafood for their primary source of protein.

8. *"We are deeply concerned that one in five people on this planet, or over 1 billion people, still live in extreme poverty, and that one in seven—or 14 percent—is undernourished..."* (Para 21. *The future we want.* UNCSD 2012)

9. Aquaculture offers huge potential for the provision of food and livelihoods, though greater efficiencies in provision of feed to aquaculture need to be realized, including reduced fish protein and oil and increased plant protein content, if the industry is to be sustainable. Aquaculture under the Blue Economy will incorporate the value of the natural capital in its development, respecting ecological parameters throughout the cycle of production, creating sustainable, decent employment and offering high value commodities for export.

10. Unsustainable Fisheries.

11. The proportion of marine fish stocks estimated to be underexploited or moderately exploited declined from 40% in the mid-1970s to 15% in 2008, and the proportion of overexploited, **depleted** or recovering stocks, increased from 10% in 1974 to 32% in 2008. Fishing fleet **subsidies** are estimated to be between US $10-30 billion per year driving the further depletion of fisheries that have otherwise ceased to be economically **viable**. The benefits lost to fishing nations as a consequence of over fishing are estimated to be in the order of $50 billion per **annum**.

12. Aquaculture is the fastest growing food sector now providing 47% of the fish for human consumption globally. The last three decades have seen massive expansion in aquaculture operations raising concerns of environmental damage and unsustainable development models. Aquaculture sites have often been carved out of important natural coastal habitats with rapid expansion exceeding the capacity of planning controls and oversight. Aquaculture with fed species, if not managed properly, can impact

biodiversity and ecosystem functions through excessive nutrient release, chemical pollution and the escape of farmed species and diseases into the natural environment.

13. It is essential that integrated ecosystem approaches are utilized in wild capture fisheries and aquaculture based on the best current scientific information with judicious application of the precautionary approach, and subsidies that encourage overfishing are removed.

14. Climate change and managing carbon budgets.

15. Sea level rise and change in ecosystem status due to changing temperatures, from coral bleaching to impacts upon migration patterns, have been discussed at length in diverse international **fora** and need not be re-stated here. Relatively new issues on the agenda, however, are Ocean Acidification and Blue Carbon.

16. Oceans are estimated to have absorbed approximately 25% of anthropogenic carbon dioxide since the commencement of the industrial revolution resulting in a 26% increase in the acidity of the Ocean. Ocean acidification is known to have a significant impact; many organisms show adverse effects, such as reduced ability to form and maintain shells and skeletons, as well as reduced survival, growth, abundance and **larval** development. Acidification will also affect carbon **accretion** in coral reef building organisms causing net decreases in global coral reef coverage and associated species. Projections suggest that pH for the more vulnerable ocean regions could reach the **aragonite** tipping point within decades changing the very chemistry of ecosystems with potentially disastrous effects. As ocean acidity increases, its capacity to absorb carbon dioxide from the atmosphere decreases, thereby reducing the ocean's capacity to moderate climate change. There is currently no international mechanism to specifically address acidification, appropriate means need to be elaborated to enable coordinated international action.

17. Several key coastal habitats such as mangroves, salt marshes and sea grass meadows have been found to fix carbon at a much higher rate per unit area than land-based systems and be more effective at the long-term sequestration of carbon than terrestrial forest ecosystems. Mangroves have been reduced to 30%–50% of their historical cover and 29% of seagrass habitats are estimated to have been lost in the last 150 years. This carbon sequestration role re-emphasizes the importance of maintaining, and where possible **rehabilitating**, such ecosystems as an opportunity for ecosystem climate mitigation.

18. The Blue Economy approach will set in place the policies, legislation, infrastructure and incentives to facilitate the transition to a low carbon economy utilizing all the

tools at its disposal including the ocean's enormous potential for renewable energy (wind, wave, tidal, thermal and biomass) generation.

19.   Marine and coastal tourism.

20.   Marine and coastal tourism is of key importance to many developing countries. Despite the global economic crisis international tourism has continued to grow. Data indicates that international tourist arrivals increased by 4% to 1. 035 billion in 2012, generating ＄1. 3 trillion in export earnings. The UNWTO forecasts further growth of 3%-4% in 2013. This does not detract however from the vulnerability of economies so heavily dependent on a single industry. Tourism brings challenges in terms of increased greenhouse gas emissions, water consumption, sewage, waste generation and loss or degradation of coastal habitat, biodiversity and ecosystem services.

21.   Pollution and marine debris.

22.   The growing human population, intensification of agriculture and the rapid urbanization of coastal areas are all key land-based factors causing higher levels of pollution in our seas. Documented marine "dead zones" now number more than 400 covering an area of over 240,000 km including some of the formerly most productive areas of **estuaries** and shelf. There has been an approximate threefold increase in the loads of nitrogen and **phosphorous** enrichment to the oceans since pre-industrial times. A recent study estimates that the "business as usual" model of nitrogen input will result in an increase of 50% in the **fluxes** of inorganic nitrogen to the Ocean by the year 2050.

23.   Sea based sources of pollution are likely to be a growing issue as maritime shipping increases and submarine hydrocarbon/mineral exploration and extraction continue to expand. Furthermore, market forces are driving exploration in ever more extreme environments posing increased ricks of marine pollution as clearly demonstrated by the 2010 Deepwater Horizon disaster in the Gulf of Mexico.

24.   Marine debris threatens the integrity of marine food chains. Plastic materials and other litter are widespread from the oceanic collection zones and **gyres**, through the **glutinous** mass of micro-plastics that can now be **trawled** from some waters to the debris and **pellets** often found in the **gastrointestinal** tract of sea and bird life.

25.   The international mechanisms in place to address these matters need implementation with renewed vigor incorporating the analysis of the true costs and benefits of rectifying these concerns in the context of the natural blue capital.

26.   Governance and international cooperation.

27. Each **sovereign** country is responsible for its own resources and sustainable development. This national responsibility and importance of national polices and development strategies should not therefore be downplayed. The principle of common but differentiated responsibilities, however, still applies. Indeed the need for structured international cooperation underpins all aspects of the Blue Economy. Whether it be with regard to updating and advancing governance mechanisms to ensure the sustainable development of waters beyond national jurisdiction (e. g. maritime security, high seas MPAs, sustainable fisheries, oil and mineral extraction) or assistance in enabling the effective management and utilization of national EEZs (e. g. technology transfer, technical assistance, marine spatial planning), capacity building, finance to support national marine spatial planning and effective monitoring, control and surveillance.

28. A key component of international cooperation for the Blue Economy approach is Research. A science-based approach is essential to the development of the Blue Economy; commencing with the initial assessment and critically the valuation of the blue capital at our disposal. This will provide a basis for informed decision-making and adaptive management. This major undertaking must be addressed and continually refined and upgraded in line with changing circumstances, evolving technologies and our increasing understanding; or the Blue Economy approach will **founder**. This underlines the importance of technical assistance, technology transfer and capacity building to the pursuit of sustainable development.

(1,668 words)

➢ *New Words*

anthropogenic [ˌænθrəpə'dʒenɪk]　　*a.* produced by human beings rather than natural forces 人为的；人类起源的

Anthropogenic carbon emissions in the atmosphere and oceans are the most significant cause of global climate change.

大气和海洋中的人为碳排放是造成全球气候变化的最主要原因。

construct [kən'strʌkt]　　*n.* an idea or a belief that is based on various pieces of evidence which are not always true 构想，概念；编造，杜撰

It is an artificial construct created by men to justify his inactions.

这是一种人造概念，是人类杜撰出来为自己的懒惰开脱的。

precursory [prɪ'kɜːs(ə)rɪ]　　*a.* warning of future misfortune 前兆的；先驱的

Animals may have the means and sensitivity to discriminate the threatening precursory signals of an impending earthquake, thus activating a behavior pattern for survival.

动物可能有某种方法，能够敏锐地觉察到地震前的危险信号，并做出某种行为以图逃生。

knock-on [ˌnɒk 'ɒn]　　*a.* causing other events to happen one after another in a series 使产生连锁反应的；有间接影响的

The plankton are the first link in many of the ocean's food chains and their demise would have a knock-on effect all the way up the food chain.

浮游生物是大多数海洋食物链的第一环节，它们的死亡将引发上游食物链的连锁反应。

mangrove ['mæŋɡrəʊv]　　*n.* a tropical tree that grows in mud or at the edge of rivers and has roots that are above ground 红树林

It has roughly 13 types of vegetation in that region, including a cloud forest, a mangrove and a freshwater herbaceous swamp.

那个地区植被大约包括 13 个树种，包括云雾林、红树林和淡水草本沼泽。

pertain [pə'teɪn]　　*v.* to exist or to apply in a particular situation or at a particular time 存在;适用

It will help if you concentrate on reading books that pertain to subjects you have an interest in.

专门阅读你感兴趣的书籍,会让你有所收获。

deplete [dɪˈpliːt]   *v.* to reduce sth. by a large amount so that there is not enough left 耗尽,用尽;使衰竭,使空虚

The current job market has also forced families to deplete their savings to pay for their everyday needs.

就业市场也不景气,所以许多家庭不得不耗尽积蓄用以支付日常开销。

subsidy [ˈsʌbsədɪ]   *n.* money that is paid by a government or an organization to reduce the costs of services or of producing goods so that their prices can be kept low 补贴;津贴;补助金

Generally, the State budget does not provide interest rate subsidy to the State Policy Banks.

一般来说,国家预算不向国家政策性银行提供利率补贴。

viable [ˈvaɪəbl]   *a.* capable of developing and surviving independently 可实施的;切实可行的;能独立发展的;能独立生存的

It is another matter whether such solar steam projects will be financially viable for their developers.

对于开发商来说,这样的太阳能蒸汽项目在经济上是否可行,那就是另外一回事了。

annum [ˈænəm]   *n.* year 年,岁

Each year Forbes. com asks a bunch of people to forecast business highlights of the new annum.

每年,福布斯网站都会请人预测新一年的商业亮点。

fora [ˈfɔːrə]   *n.* (forum 的复数形式) a place where people can exchange opinions and ideas on a particular issue 论坛;讨论会;法庭;专题讲话节目

We need to find ways to move forward together at these fora.

我们需要通过这些论坛寻求共同发展的道路。

larval [ˈlɑːvl]   *a.* immature of its kind; especially being or characteristic of immature insects in the newly hatched wormlike feeding stage 幼虫的;幼虫状态的

Dragonflies spend their larval stages in water, and therefore aquatic environment is vital for their breeding cycles.

蜻蜓在水中渡过幼虫阶段,因此水生环境对它们的繁殖周期非常关键。

accretion [əˈkriːʃn]	*n.*	the process of new layers being slowly added to sth. 堆积,积聚(过程);积聚层;堆积层 A coral reef is built by the accretion of tiny, identical organisms. 珊瑚礁是由许多相同的微生物堆积而成的。
aragonite [ˈærəɡ(ə)naɪt]	*n.*	a mineral form of crystalline calcium carbonate 霰石;文石 Some corals have skeletons composed of aragonite, a particularly unstable form of calcium carbonate. 有些珊瑚的骨架中含有霰石,它是一种极不稳定的碳酸钙形式。
rehabilitate [ˌriːəˈbɪlɪteɪt]	*v.*	to return a building or an area to its previous good condition 使(建筑物或地区)恢复原状;修复 The city still needs to rebuild and rehabilitate buildings that were destroyed by the floods. 这个城市仍然需要重建和修复被洪水摧毁的建筑。
estuary [ˈestʃuərɪ]	*n.*	the wide part of a river where it nears the sea 河口;江口 The estuary is an important habitat for birds; and large barrages would damage much of it, as well as interfere with fish stocks in the river. 河口是鸟类的重要栖息地,在此建立大型水坝,会对其造成大规模破坏,还会对鱼群的生态环境造成影响。
phosphorous [ˈfɒsf(ə)rəs]	*a.*	containing or characteristic of phosphorus 磷的,含磷的;三价磷的;发磷光的 Livestock are estimated to be the main inland source of phosphorous and nitrogen contamination of the South Sea, contributing to biodiversity loss in Marine ecosystems. 据估计,畜牧业是南海磷污染和氮污染的主要内陆来源,造成海洋生态系统生物多样性的缺失。
flux [flʌks]	*n.*	an act of flowing 不断的变动;不停的变化 Scientists aren't sure if the world's deserts collectively are growing, but the boundaries of some are in flux. 科学家们不确定全球沙漠面积是否正在扩大,但有些沙漠的边界确实是在不断变化。
gyre [ˈdʒaɪə]	*n.*	a round shape formed by a series of concentric circles

涡旋,环流;旋转,回旋

In the broad expanse of the northern Pacific Ocean, there exists the north Pacific Subtropical Gyre, a slowly moving, clockwise spiral of currents.

在广阔的北太平洋海域中,有一股北太平洋副热带环流,正在以顺时针方向螺旋式缓缓移动。

glutinous ['gluːtənəs]   *a.* having the sticky properties of an adhesive 黏的,黏性的;胶状的,胶质的

Analysts say the prices of raw materials of rice dumplings, including glutinous rice, sugar and edible oil, have risen 25% this year, causing a spike in rice dumpling price.

专家分析说,今年粽子的原料如糯米、糖、油等价格上涨了25%,粽子价格也因此水涨船高。

trawl [trɔːl]   *v.* to fish for something by pulling a large net with a wide opening through the water 用拖网捕鱼

She would walk on to the beach and watch the night fishermen trawl the shallow waters.

她会步行到海滩去看夜间工作的渔民在浅水中拖网捕鱼。

pellet ['pelɪt]   *n.* a small sphere 颗粒;小球

In order to reduce the diesel particulate matter, the pellet filter is installed on the automobile.

为减少柴油燃烧微粒排放,汽车上安装了颗粒过滤器。

gastrointestinal [ˌgæstrəʊɪn'testɪnl]   *a.* of or relating to the stomach and intestines 胃肠的

In early childhood, allergy is frequently associated with gastrointestinal dysfunction.

儿童过敏常常和胃肠道功能障碍有关。

sovereign ['sɒvrɪn]   *a.* free to govern itself; completely independent 主权的;至高无上的

A strike against a sovereign nation raises moral and legal issues.

对一个主权国家发动攻击会引起道德和法律上的争议。

founder ['faʊndə(r)]   *v.* to fail because of a particular problem or difficulty 失败;破产

The talks have foundered, largely because of the reluctance of some members of the government to do a deal with criminals.

谈判已经破裂,主要是因为一些政府成员不愿意和犯罪分子达成协议。

➤ *Phrases and Expressions*

be made up of 由······所组成

in point 相关的;恰当的;中肯的

factor in 将······纳入;把······计算在内

be lost to 不再属于······所有

in the order of 大约

tipping point 临界点;引爆点

at one's disposal 由其支配;由某人做主

in line with 符合;与······一致

➤ *Terminology*

coral bleaching 珊瑚白化,珊瑚颜色变白的现象,珊瑚本身是白色,其美丽颜色来自体内的共生海藻,珊瑚依赖体内的共生海藻生存,海藻通过光合作用向珊瑚提供能量。如果共生海藻离开或死亡,珊瑚就会变白,最终因失去营养供应而死。由于海洋温度不断升高,致使珊瑚所依赖的海藻减少,珊瑚也因此更易受到白化的影响。三十年前,大规模的珊瑚白化现象比较罕见,但近年来却越来越多地出现

pH 酸碱度

salt marsh 盐沼

sea grass meadows 海草床,大面积的连片海草

dead zone (海洋)死区,因海水严重富营养化而造成的鱼类等生物无法生存的区域

inorganic nitrogen 无机氮

➤ *Proper Names*

principle of common but differentiated responsibilities 共同但有区别责任原则,由于地球生态系统的整体性和在导致全球环境退化过程中发达国家和发展中国家的不同作用,各国对保护全球环境应负共同但有区别的责任。它包括两个方面,即共同的责任和有区别的责任。

The Sunken Billions 沉没的数十亿,世界银行与联合国粮食及农业组织在 2008 年联合进行的一项研究,发现对渔业资源的不当管理造成每年大约 500 亿美元的损失。

MPA (Marine Protected Area) 海洋保护区

CBD (Convention on Biological Diversity) 生物多样性公约

WSSD (World Summit on Sustainable Development) 世界可持续发展公约

Ocean Acidification 海洋酸化

Blue Carbon 蓝碳,利用海洋活动及海洋生物吸收大气中的二氧化碳,并将其固定、储存在海洋

中的过程、活动和机制,海草床、红树林、盐沼被认为是 3 个重要的海岸带蓝碳生态系统,研究表明,大型海藻、贝类乃至微型生物也能高效固定并储存碳。

UNWTO（United Nations World Tourism Organization）联合国世界旅游组织

Deepwater Horizon 文中指"深海地平线"钻井平台漏油事件,"深海地平线"是位于美国路易斯安那州海岸约 80 千米的一座近海钻井油田,2010 年 4 月 20 日发生爆炸并沉没,导致大规模的墨西哥湾原油污染。

Gulf of Mexico 墨西哥湾

➤ *Translation*

1. The natural world made up of the physical environment, its mineral... （Para. 1）

我们生活的自然世界,包括物理环境及其矿物成分,还有遗传、物种和生态系统三个层面的生物多样性,本身就有千丝万缕的联系。这个自然体系越多样,生命力越强,这种联系就越紧密。

2. An ecosystem approach is required that factors in restoration of... （Para. 5）

因此,现在需要一种生态体系的思路,考虑如何恢复生物多样性与可再生资源以及如何科学管理资源开发活动。

3. In the context of the Blue Economy food security is very closely... （Para. 7）

在蓝色经济理念体系中,食品安全与生物多样性的可持续利用息息相关,尤其是在野生渔业资源开发相关领域。

4. The last three decades have seen massive expansion in aquaculture... （Para. 12）

在过去的三十年中,水产养殖业得以大规模发展,这使人们开始担忧对环境造成的破坏以及这种毁灭式发展所带来的问题。

【英语中经常使用无灵主语,意在使叙述显得客观、公正,语气较为委婉、间接,如原文中使用"last three decades"作为"see"的主语,但是汉语中不能这样搭配,因此在译文中将其译为时间状语,将来源于动词的名词"expansion"转译为动词,作为谓语,选择"aquaculture operations"作为主语,将原文中的后置定语"raising..."从原句中拆分出来,并增译虚指主语"人们",并根据语义增译"所带来的问题"。】

5. It is essential that integrated ecosystem approaches are utilized in... （Para. 13）

在野生捕捞和水产养殖行业,根据最新的科学信息,采取明智的预防措施,使用综合的生态方法,同时取消渔业补贴,以免导致过度捕捞,这些做法都是非常必要的。

6. Sea level rise and change in ecosystem status due to changing... （Para. 15）

温度变化导致海平面上升,生态系统状态也发生变化,从引发珊瑚白化到影响鱼类洄游,种种问题都已在各种国际论坛中进行了详尽的讨论,此处无须赘述。

7. Ocean acidification is known to have a significant impact; many... （Para. 16）

众所周知,海洋酸化影响巨大,其负面影响在许多物种身上都已有所体现,例如会妨碍贝壳类生物长出外壳,即使长出也容易受到侵蚀,还会影响其他海洋生物的骨骼发育,造成海洋生物存活率降低,个体成长、种群丰度和幼体发育能力都会受到影响。

【首先将"is known"拆分独立,再根据上下文语义以及专业背景知识对原文中的"form and maintain"进行扩充,增译如"贝壳类生物""也容易受到侵蚀""影响其他海洋生物的……发育",有利于目的语读者理解。】

8. This does not detract however from the vulnerability of economies...（Para. 20）

但是,如果一个国家的经济严重依赖某种单一行业的话,那么其经济体系将是非常脆弱的,即便是旅游业持续增长,也不会改变其脆弱本质。

【原文是典型的树形结构,翻译时需将其拆分为竹式结构,将"vulnerability"转译为形容词,将作为后置定语的形容词词组"dependent on a single industry"扩充为一句话,从原文中拆分出来作为状语从句,增译"如果……那么……"体现各个分句之间的逻辑关系,同时根据上下文将"this"译为"即便是旅游业持续增长",以方便目的语读者理解。】

9. The growing human population, intensification of agriculture and...（Para. 22）

人口不断增长,农业集约化程度不断提高,沿海地区加快城市化进程,所有这些虽然是在陆地上发生的事情,但却是加重海洋污染的关键因素。

（农业集约化是在一定面积的土地上投入较多的劳动、资金和技术,以期取得较高的单位面积产量,同时减少每单位产品劳动耗费的一种农业经营方式。）

10. Furthermore, market forces are driving exploration in ever more...（Para. 23）

此外,在市场需求的驱动下,人们开始向更远处深入进行勘探,这也增加了海洋污染的风险,2010 年墨西哥湾"深海地平线"事件就是明证。

（"深海地平线"是位于美国路易斯安那州海岸约 80 千米的一座近海钻井油田,2010 年 4 月 20 日发生爆炸并沉没,导致大规模的墨西哥湾原油污染。）

11. This major undertaking must be addressed and continually refined...（Para. 28）

这是一项重大任务,必须予以重视,而且还要根据情况的变化,持续深入、不断完善,推动技术发展,深化我们的理解,否则"蓝色经济"构想及其方法将会一败涂地。

 **Exercises**

1. Fill in the blanks with the proper given words, and then translate the sentences into Chinese.

intrinsically	in point	exploitation	equivalent
renowned	determination	moderately	capture
survival	abundance	vulnerable	tipping point
elaborate	facilitate	at one's disposal	in line with

1) Research suggests that the IQ drop caused by electronic obsession is _____ to a wakeful night.

2) The increasing _____ of resources threatens to exhaust or unalterably spoils forests, soils, water, air and climate.

3) It will be better if we can learn to grasp communication skills as they can _____ mutual understanding.

4) The regulations allow countries to take measures _____ their own national conditions, citing airport screening and quarantine as acceptable, for instance.

5) When we do something for its own sake because we enjoy it or because it fills some deep-seated desire, we are _____ motivated.

6) In many cases religious persecution is the cause of people fleeing their country and a case _____ is colonial India.

7) We didn't bother to find a hotel, for my good friend invited us to stay in his home and has a luxurious car _____.

8) Located at the mouth of the Mississippi River, this city is _____ for its jazz music, wild nightlife and cuisine.

9) The factors that cause a charged crowd to reach a _____ and erupt into violence are not well understood by scientists because crowd behavior is so difficult to study.

10) They also said that the study subjects only _____ lowered their sodium intake, so the effect on blood pressure and heart disease was small.

11) It is true that everyone has a desire for success, but not everyone has the courage and _____ to pursue it.

12) With the changes in the world's climate, dinosaurs died, but many smaller animals lived on; and this was the _____ of the fittest.

13) Pictures that _____ a moment in time can evoke all sorts of feelings when looked at later.

14) Babies lose heat much faster than adults, and are especially _____ to the cold in their first

month.

15) What is really important is the awareness of the _____ of happiness and joy we have right now, within ourselves.

16) You can _____ on what you see and write it in your personal style without missing anything.

2. Translate the following passage into Chinese.

Each sovereign country is responsible for its own resources and sustainable development. This national responsibility and importance of national polices and development strategies should not therefore be downplayed. The principle of common but differentiated responsibilities, however, still applies. Indeed the need for structured international cooperation underpins all aspects of the Blue Economy. Whether it be with regard to updating and advancing governance mechanisms to ensure the sustainable development of waters beyond national jurisdiction (e. g. maritime security, high seas MPAs, sustainable fisheries, oil and mineral extraction) or assistance in enabling the effective management and utilization of national EEZs (e. g. technology transfer, technical assistance, marine spatial planning), capacity building, finance to support national marine spatial planning and effective monitoring, control and surveillance.

3. Translate the following passage into English.

一直以来,我们都致力于保护陆地濒危物种,其实对待海洋资源也应该一样。然而,为了获取经济利益,我们对海洋资源进行了过度开发,现在是时候采取根本性的措施来扭转这种局面了,我们要把重心转到海洋保护工作上来。目前,鱼类资源逐渐减少,而捕捞配额却没有减少,这就造成了一种恶性循环,我们应该彻底打破这种恶性循环,要建立海洋保护区禁止渔猎,只有这样才能拯救那些濒临灭绝的鱼类,要保护海洋生态系统,合理开发、利用海洋资源,要切实保护海洋资源,让其得以休养生息。

Unit 3
Opportunities and Investment

The Blue Economy—Opportunities

1. Issues and problems bring with them challenges and opportunities and the Blue Economy offers a suite of opportunities for sustainable, clean, **equitable** blue growth in both traditional and emerging sectors.

2. Shipping and Port **Facilities**

3. 80 percent of global trade by volume, and over 70 per cent by value, is carried by sea and handled by ports worldwide. For developing countries, on a national basis, these percentages are typically higher. World seaborne trade grew by 4% in 2011, to 8. 7 billion tons despite the global economic crisis and container traffic is projected to triple by 2030. Coastal countries and SIDS need to position themselves in terms of facilities and capacities to cater for this growing trade and optimize their benefits. The IMO has brought in new industry wide measures to increase efficiency, reduce greenhouse gas emissions and pollution. More needs to be done to address the issues of IAS from ballast water and hull fouling but even with these challenges maritime trade is set fair for growth and economic benefits whilst reducing impacts, offering expanding Blue employment opportunities for the foreseeable future.

4. Fisheries Globally

5. 350 million jobs are linked to marine fisheries, with 90% of fishers living in developing countries. The value of fish traded by developing countries is estimated at $ 25 billion, making it their largest single trade item. Global catch rose from 4 million tons in 1900, through 16. 7 million tons in 1950, 62 million tons in 1980 to 86. 7 million tons in 2000 but has **stagnated** subsequently. In 2009 marine capture production was 79 million tons. Overall catch risks decline with 75% of stocks fully exploited or depleted. Human activity has directly and markedly reduced ocean productivity; additional **deficits** may be due to climate change increasing ocean **stratification** and reducing nutri-

ent mixing in the open seas. Global Ocean Observing System (GOOS) and LME assessments show significant warming trends from which model projections 2040-2060 forecast a steady decline in ocean productivity.

6.　The implementation of integrated, ecosystem-based approaches based on the best available science in a **precautionary** context, plus the removal of fishery subsidies that drive overexploitation offer the prospect of restoring key stocks and increasing catches. It is estimated that $ 50 billion per annum is lost to overfishing and could be progressively recovered through stock restoration. The implementation of sound management measures brings the promise of increased sustainable catches, lower energy utilization and costs; thereby securing livelihoods and enhancing food security.

7.　Tourism

8.　Tourism is a major global industry; in 2012 international tourist arrivals increased by 4% despite the global economic crisis and constituted 9% of Global GDP (direct, indirect and induced impact). In 2012 tourism supported 9% of global jobs and generated $ 1. 3 trillion or 6% of the world's export earnings. International tourism has grown from 25 million in 1950 to 1,035 million in 2012 and the UNWTO forecasts further growth of 3% –4% in 2013; the forecast for 2030 being 1. 8 billion. A large portion of global tourism is focused on the marine and coastal environment and it is set to rise. Trends in aging populations, rising incomes and relatively low transport costs will make coastal and ocean locations ever more attractive. Cruise tourism is the fastest growing sector in the leisure travel industry; between 1970 and 2005 the number of passengers increased 24-fold to 16 million by 2011. Overall, average annual passenger growth rates are in the region of 7. 5% and passenger **expenditures** are estimated in the order of $ 18 billion per year.

9.　Tourism developments bring various problems. The tourism consumer, however, is driving the transformation of the sector with a 20% annual growth rate in ecotourism; about 6 times the rate of growth of the overall industry. A Blue Economy approach where ecosystem services are properly valued and incorporated into development planning will further advance this transition, guiding tourism development and promoting lower impact activities, such as ecotourism and nature-based tourism, where the natural capital is maintained as an integral part of the process.

10.　Aquaculture

11.　Aquaculture is the fastest growing global food sector now providing 47% of the fish for human consumption. Fish used for human consumption grew by more than 90 million tons in the period 1960—2009 (from 27 to 118 million tons) and aquaculture is projected to soon **surpass** capture fisheries as the primary provider of such protein.

12. To maintain its viability and growth without undermining wild fisheries the aquaculture industry must actively reduce the proportion of industrial fish in **fishmeal**. Progress is being made however; fishmeal is increasingly being produced from fishery by-products, which now constitute over 25% of global production. Research indicates that at least 50% of fishmeal and 50% – 80% of oil in **salmonid** (the largest component of aquaculture production) and 30% – 80% of fishmeal and up to 60% of oil in marine fish diets can ultimately be replaced with vegetable **substitutes**, greatly increasing the scope for industry expansion.

13. Energy

14. In 2009 offshore fields accounted for 32% of worldwide crude oil production and this is projected to rise to 34% in 2025 and higher subsequently, as almost half the remaining recoverable **conventional** oil is estimated to be in offshore fields-a quarter of that in deep water. Deep water oil drilling is not new, but market pressures are making the exploration for and tapping of evermore remote **reserves** cost effective, bringing the most isolated areas under consideration. Methane hydrates, a potentially enormous source of hydrocarbons, are now also being explored and tapped from the seabed.

15. Oil will remain the dominant energy source for many decades to come but the Ocean offers enormous potential for the generation of renewable energy-wind, wave, tidal, biomass, thermal conversion and **salinity gradients**. Of these the offshore wind energy industry is the most developed of the ocean-based energy sources. Global installed capacity was only a little over 6 GW in 2012 but this is set to quadruple by 2014 and relatively conservative estimates suggest this could grow to 175 GW by 2035.

16. Biotechnology

17. The global market for marine biotechnology products and processes is currently estimated at $2.8 billion and projected to grow to around $4.6 billion by 2017. Marine biotech has the potential to address a suite of global challenges such as sustainable food supplies, human health, energy security and environmental **remediation**. Marine **bacteria** are a rich source of potential drugs. In 2011 there were over 36 marine derived drugs in clinical development, including 15 for the treatment of cancer. One area where marine biotech may make a critical contribution is the development of new **antibiotics**. The potential scope is enormous, by 2006 more than 14,000 novel chemicals had been identified by marine bioprospecting and 300 **patents** registered

on marine natural products.

18. On the energy front **algal** biofuels offer promising prospects. The European Science Foundation **postulates** a production volume of 20–80 thousand liters of oil per **hectare** per year can be achieved from microalgal culture, with even the lower part of this range being considerably higher than terrestrial biofuel crops.

19. Submarine mining

20. The world is **gearing** up for the exploration and exploitation of mineral **deposits** on and beneath the sea floor. Industry, due to rising commodity prices, is turning its attention to the potential riches of polymetallic nodules, cobalt crusts and massive sulphide deposits; the latter a source of rare earth elements, such as yttrium, dysprosium and terbium, important in new ICT hardware and renewable energy technologies. Commercial interest is particularly strong in polymetallic nodules and in seafloor massive sulphides.

21. The International Seabed Authority has developed the Mining Code regulations to meet these changing circumstances and has commenced issuing licenses for the exploration of the international sea floor. Coastal countries need to prepare themselves to ensure they realize optimal benefits from resources in their own EEZs and likewise that their concerns are incorporated into the measures to manage the coming race for the riches of the seabed.

22. To realize the necessary international cooperation and support to elevate the Blue Economy to the international sustainable development agenda, diplomatic effort amongst SIDS has targeted the preparatory process leading up to the Third International Conference on Small Island Developing States in Apia, Samoa 2014.

23. It is proposed to hold a "Blue Summit" in Abu Dhabi in January 2014, as part of the Sustainable Development Week with the support of the Government of the United Arab Emirates. This will allow SIDS groupings and other partners (e. g. IOR-ARC, etc.) to contribute to an internationally endorsed Blue Economy document for **submission** to the Third International Conference on Small Island Developing States in Samoa, 2014. Full development and endorsement of the proposal in Samoa would constitute the next step in securing international momentum for, and acceptance of, the Blue Economy as an approach distinct from, but mutually supportive with, the Green Economy model.

(1,475 words)

➤ *New Words*

equitable ['ekwɪtəbl] *a.* fair and reasonable; treating everyone in an equal way 公平的,公正的

WHO considers equitable access to safe and affordable medicines as vital to the attainment of the highest possible standard of health by all.

世卫组织认为,在安全用药方面实现公平,保证人人买得到、买得起,这是实现全民健康最高标准的关键。

facility [fə'sɪlətɪ] *n.* a building or place that provides a particular service or is used for a particular industry 设备;设施

We received a number of complaints from customers about the lack of parking facilities.

我们接到很多顾客投诉说停车设施太少了。

stagnate [stæg'neɪt] *v.* to stop developing or making progress 使停滞;使萧条;使淤塞

Whether you're looking at your job, your relationship or your physical state of being, when you slip into comfortable territory, you'll inevitably stagnate.

不论是工作,还是人际关系,或者身体状态,一旦陷入舒适区太久,就会不可避免地停滞不前。

deficit ['defɪsɪt] *n.* the property of being an amount by which something is less than expected or required 赤字;不足额

The United States Congress and the President are still locked in disagreement over proposals to reduce the massive budget deficit.

美国国会和总统尚未就削减巨额预算赤字的提案达成一致,双方仍然僵持不下。

stratification [ˌstrætɪfɪ'keɪʃn] *n.* the division of something, especially society, into different classes or layers 分层;成层

Many have noted the increasing stratification of American society and that, unlike in decades past, entry into its top levels now depends largely on graduation from elite universities.

许多人注意到,美国社会的阶层分化日益加剧,而且与过去几十年不同的是,现在要想进入上流社会,很大程度上取决于是否从精英大学毕业。

precautionary [prɪ'kɔːʃənərɪ] *a.* taken in advance to protect against possible danger or failure

预防的;留心的;预先警戒的

Police said they planned to add 40 percent more officers as a precautionary measure.

警方表示,作为预防措施,他们准备增加40%的警力。

expenditure [ɪkˈspendɪtʃə(r)]　*n.* money paid out 支出,花费;经费,消费额

Policies of tax reduction must lead to reduced public expenditure.

减税政策必然导致公共开支减少。

surpass [səˈpɑːs]　*v.* to do or be better than sb./sth. 超越;胜过,优于

By the end of this year, the number of tablet and smartphone game players will surpass one billion.

到今年年底,平板电脑和智能手机的游戏玩家数量将超过十亿。

fishmeal [ˈfɪʃmiːl]　*n.* ground dried fish used as feed for farm animals, as a fertilizer, etc. (用作饲料或肥料的)鱼粉

Fishmeal is an important animal protein feed, which is rich in protein and vitamins and is a major raw material of animal feed.

鱼粉作为重要的蛋白质饲料,其蛋白质含量高且具有丰富的维生素,是主要的饲料原料。

salmonid [ˈsælmənɪd]　*n.* soft-finned fishes of cold and temperate waters 鲑科鱼

Tilapia, the third largest culture species in the world after Carps and Salmonids, is one of the most important freshwater culture species in China.

罗非鱼是世界上第三大养殖品种,仅次于鲤科和鲑科,也是我国最重要的淡水鱼养殖品种之一。

substitute [ˈsʌbstɪtjuːt]　*n.* a person or thing that takes or can take the place of another 代用品;代替者

The printed word is no substitute for personal discussion with a great thinker.

与伟大的思想家面对面交流,这种体验是任何文字都无法代替的。

conventional [kənˈvenʃənl]　*a.* following what is traditional or the way something has been done for a long time 符合习俗的,传统的;常见的;惯例的

It became difficult to promote conventional ideas of excellence without being instantly accused of elitism.

一倡导传统优秀价值观,就会被指责是精英主义,这是很难避免的。

reserve [riˈzɜːv]　*n.* a supply of something that is available for use when it is needed 储备(量);储藏(量);(动植物)保护区;自然保护区

To harvest oil and gas profitably from the North Sea, we must focus on the exploitation of small reserves as the big wells are running dry.

大油井正濒临枯竭,要想从北海开采石油和天然气并从中获利,就必须集中精力开采小储量油井。

salinity [səˈlɪnətɪ] *n.* the relative proportion of salt in a solution 盐度;盐分;盐性

The fear is that melting ice, along with increased snow and rain, could reduce the density and salinity of the top layers of the sea, making them more buoyant.

人们担忧的是冰层融化、雨雪量增加,这些可能会导致海洋表层水体的密度和盐度降低,如此一来,表层水体就更容易滞留在海洋表层。

gradient [ˈɡreɪdɪənt] *n.* the rate at which temperature, pressure, etc. changes or increases, between one region and another(温度、气压等的)变化率,梯度变化曲线

Prevalence was 5.1 times as high in men as in women, increased with age, was higher in rural than in urban or remote areas and showed a north-to-south gradient.

男性的患病率是女性的 5.1 倍,年龄越大患病率越高,农村地区的患病率高于城市或偏远地区,从北向南呈上升趋势。

remediation [rɪˌmiːdɪˈeɪʃn] *n.* act of correcting an error or a fault or an evil 修复;补救;矫正;补习

Microbial degradation is one of the economical and effective methods for the remediation of petroleum pollution.

针对石油污染,采取微生物降解是一种经济有效的补救措施。

bacteria [bækˈtɪərɪə] *n.* very small organisms, some can cause disease 细菌

Parts of the shop were very dirty, unhygienic, and a breeding ground for bacteria.

商店里有些地方非常脏而且不卫生,是细菌滋生之地。

antibiotic [ˌæntɪbaɪˈɒtɪk] *n.* a chemical substance derivable from a mold or bacterium that kills microorganisms and cures infections 抗生素

Antibiotic-resistant bacteria infect 100,000 patients a year, and they are notoriously hard to fight.

耐抗生素细菌每年感染 10 万名患者,而且它们是出了名的难以对付。

patent [ˈpæt(ə)nt] *n.* an official right to be the only person to make, use or sell a product or an invention 专利;专利权;专利证;专利品

Traditionally, financial inventions were not granted patent rights, but now in the United States and in a number of other countries it has become possible to patent them.

传统来讲,金融产品是不能获批专利的,但现在,在美国和其他的一些国家,金融产品也可能申请专利。

algal ['ælgəl]

a. of or relating to alga 海藻的

Such algal blooms are not only harmful to people and animals but can also trigger dead zones in the lake-areas in the water so devoid of oxygen that they cannot support aerobic life.

这种水藻爆发不仅会对人畜造成危害,还会导致湖区出现死亡区域,也就是说,水中某片区域会因氧气过度匮乏而无法满足有氧生物存活需要。

postulate ['pɒstʃəleɪt]

v. to suggest or accept that something is true so that it can be used as the basis for a theory, etc. 假定;假设

Scientists do not know why dinosaurs became extinct, but some theories postulate that changes in geography, climate and sea levels were responsible.

科学家们还无法确定恐龙为何灭绝,但有些理论推断这与地理气候条件和海平面的某些变化有关。

hectare ['hekteə(r)]

n. a unit for measuring an area of land; 10,000 square meters or about 2.5 acres 公顷(等于1万平方米)

China produces twice as much rice per hectare as India with the same volume of water.

在同样的水量下,中国每公顷的水稻产量是印度的两倍。

gear [gɪə(r)]

v. to prepare yourself/sb./sth to do sth. 使……准备好;开动;搭上齿轮;使……适合

It will be a year of manipulation and mischief, as politicians of all stripes gear up for the 2012 presidential election.

2012年总统竞选来临之际,政界官员们无论级别大小,都蠢蠢欲动,又是暗箱操作和钩心斗角的一年。

deposit [dɪ'pɒzɪt]

n. a layer of a substance that has formed naturally underground 沉淀物

In the world of seabed geology, if a deposit is massive, it is not necessarily big, but rich in metals.

按海底地质界的说法,如果说某个矿床巨大,并不一定是面积或者体积大,而是说其金属含量高。

submission [səb'mɪʃn]

n. the act of giving a document, proposal, etc. to sb. in authority so that they can study or consider it 提交；呈递；提交（或呈递）的文件、建议等

No date has yet been set for the submission of applications. 申请的提交日期还未确定。

➤ *Phrases and Expressions*

a suite of 一系列

cater for 迎合；为……提供所需；供应伙食

set fair 天气稳定晴朗

a large portion of 大部分

under consideration 在考虑之中

gear up for 为……做准备

lead up to 导致；作为……的准备

➤ *Terminology*

greenhouse gas emission 温室气体排放

hull fouling 船体生物污损

ecotourism 生态旅游

methane hydrate 甲烷水合物

global installed capacity 全球装机容量

GW 全称 Gigawatt, 十亿瓦特

microalgal culture 微藻养殖

polymetallic nodules 多金属结核

cobalt crusts 富钴结壳，生长在海底岩石或岩屑表面的皮壳状铁锰氧化物和氢氧化物，富含钴

massive sulphide deposit 块状硫化物

yttrium 钇

dysprosium 镝

terbium 铽

➤ *Proper Names*

IAS（Invasive Alien Species）外来物种入侵

GOOS（Global Ocean Observing System）全球海洋观测系统

LME（Large Marine Ecosystem）大洋生态系统

European Science Foundation 欧洲科学基金会

ICT（Information and Communication Technology）信息通信技术

International Seabed Authority 国际海底管理局

Third International Conference on Small Island Developing States 第三届小岛屿发展中国家国际

会议
Blue Summit 蓝色峰会
Abu Dhabi 阿布扎比
Sustainable Development Week 可持续发展周
IOR-ARC 环印度洋区域合作联盟,也称环印度洋地区合作联盟,简称环印联盟

➢ *Translation*

1. Issues and problems bring with them challenges and opportunities...（Para. 1）
   争议和问题也带来了挑战与机遇,不管是传统行业,还是新兴领域,蓝色经济都带来了一系列发展机会,在保证环保、公平的前提下,实现可持续的蓝色增长。

2. More needs to be done to address the issues of IAS from ballast...（Para. 3）
   目前,压舱水和船体生物污损带来的外来物种入侵问题还亟待解决,但是,即便面对这么多困难,海上贸易前景依然乐观,有望在降低环境影响的同时带来不小的经济效益,在不远的将来还会提供更多的蓝色就业机会。
   （压舱水是为了保持船舶平衡而专门注入压载水舱的水,是船舶安全航行的重要保证,特别是对没有装载适量货物的船舶,适量压舱水可保证船舶的螺旋桨吃水充分,将船舶尾波引发的船体震动降低到最低限度,并维持推进效率。但同时,压舱水也是外来海洋生物入侵的重要载体,每天通过船体及压舱水在世界各地转运的生物物种数可能超过 3 000 种。船体生物污损是船体上各种水生生物的不良积累,所有船舶都会受到一定程度的影响,污垢类型包括细菌和硅藻,如藤壶、管虫和藻类,这些植物和动物将被船舶带到各个海域,如果它们能够幸存下来,就有可能建立一个生物种群,与本地物种产生竞争。通过全球航运活动转移导致的外来物种入侵会对环境产生不利影响或者破坏海洋资源,被认为是对全球海洋和沿海生态系统的重大威胁,越来越引起立法者和航运界的关注。）
   【该句难点来自"IAS"的翻译,查询所有网络词典以及外文搜索引擎,使用最为普遍的义项都是"国际会计准则",但从上下文语境可以明显看出该义项不适用于本文,再次查询,发现另一个没有被字典收录的义项"国际认证服务,是国际规范委员会（International Code Council,简称 ICC）的附属机构,为包括政府机构、商业企业和专业协会在内的各种企业和组织提供认证服务,目前认证范围主要集中在公共安全和可持续发展项目",该义项和上下文有一定关联,但依然难以理顺,经查询大量专业资料,确定"IAS"全称为"invasive alien species",即"外来物种入侵",才是最适合本文的义项。从本例可以看出,两种语言的不完全对应现象给翻译带来了很大困难,尤其目的语中有多个选择的时候,一定要根据上下文以及专业背景来确定词义,仔细求证,交叉验证。】

3. The implementation of sound management measures brings...（Para. 6）
   健全制度、完善措施,能够保证提高可持续捕捞量,降低能耗和成本,从而保障民生、提高食品安全。
   【原文集中体现了英语树形结构和静态语势的特点,句子主干只有一个谓语动词"bring",分支上有很多来源于动词的名词或者形容词,所以翻译时的主要思路就是拆分成竹式结构,并且利用转译来体现汉语的动态语势。首先,将"implementation"转译为动词,将

"sound"也转译为动词,组成一个并列结构;其次,省译原文谓语"bring",同时将"promise"转译为动词,代其成为谓语;另外,将"increased"和"lower"都转译为动词,组成一个并列结构。】

4. Trends in aging populations, rising incomes and relatively low... (Para. 8)

老龄群体越来越大,人们收入越来越高,(所以旅游需求越来越多),而水路运输成本相对较低,因此,沿海和海洋地区旅游线路越来越受到游客的青睐。

5. To maintain its viability and growth without undermining wild... (Para. 12)

要维护水产养殖业的生存及发展,同时还不能破坏野生渔业,这就要求水产养殖业必须积极引入新型鱼用饲料,以减少鱼粉中使用工业捕捞用鱼的比例。

(鱼粉是用一种或多种鱼类为原料,经去油、脱水、粉碎加工后的高蛋白质饲料原料,是重要的动物性蛋白质添加饲料,目前正在研究低鱼粉日粮和无鱼粉日粮,用低廉而丰富的动植物蛋白源及单细胞蛋白源部分或完全代替鱼粉。)

6. Deep water oil drilling is not new, but market pressures are making... (Para. 14)

深水石油钻探并不是什么新鲜事物,但是市场强调成本效益,这就使我们不得不把目光投向更偏远的地区,在那里进行勘探和开发。

7. One area where marine biotech may make a critical contribution is... (Para. 17)

海洋生物技术有可能会在开发新型抗生素领域大有作为。

8. The world is gearing up for the exploration and exploitation of... (Para. 20)

全球都在摩拳擦掌,准备开始海底勘探采矿。

9. Coastal countries need to prepare themselves to ensure they realize... (Para. 21)

沿海国家需要做好准备,以确保自己能从专属经济区资源开发中获得最大收益,同时,也应确保在这场海底资源开发竞赛中,与各国一起议定相应管理措施的时候,一定要将自己国家的利益和关切纳入考虑范围。

10. Full development and endorsement of the proposal in Samoa... (Para. 23)

如果提交到萨摩亚的议案能够得以完善并赢得支持,那么下一步将会在国际上掀起一股蓝色经济的潮流,推动国际社会接受蓝色经济的理念,这种理念与之前的绿色经济不同,但与其相辅相成。

(2014年9月1日,第三届小岛屿发展中国家国际会议在南太平洋岛国萨摩亚首都阿皮亚举行,会议主题为"通过真正与持久的合作伙伴关系促进小岛屿发展中国家的可持续发展",来自全球数十个小岛屿发展中国家的代表齐聚一堂,探讨如何通过可持续发展应对小岛屿国家所面临的挑战,会议最终通过名为《小岛屿发展中国家快速行动方式》的成果文件。)

【与例3一样,本句也是集中体现了英语树形结构的特点,句子很长,却是一个简单句,只有一个谓语动词"constitute",其他句子成分以各种形式附着在主干上,针对这种情况,主要翻译思路就是拆分。首先,为"development and endorsement"增译动词"得以",从而将这个作为主语的名词词组从原文中拆分出来,单独成句;其次,省译"constitute",增译"掀起""推动"这种动作性比较强的动词,同时将"acceptance"也转译为动词;另外,将作为后置定语的形容词词组"distinct from"从原文中拆分出来,与其修饰的名词"approach"分别作为主语和谓语,单独组句。】

 **Exercises**

1. Fill in the blanks with the proper given words, and then translate the sentences into Chinese.

facility	project	cater for	set fair
a large portion of	expenditure	approach	substitute
conventional	remote	under consideration	conservative
clinical	gear up for	issue	lead up to

1) We chose a hotel which is not a hotel in the _____ sense, but rather a whole village turned into a hotel.

2) What I want to remind you is that there's no _____ for practical experience although the course teaches you the theory.

3) It is _____ to create 2 million jobs over the next two years, half of the 4 million total envisaged by the program.

4) He said the authorities were trying to find shelter for the passengers at nearby schools, conference halls and other public _____.

5) The Geopark will bring real and sustainable development to the very _____ and economically deprived area.

6) _____ evidence began to accumulate, suggesting that the new drugs had a wider range of use.

7) The different mode of thinking between the Western people and Chinese people _____ the different form of expression between English and Chinese.

8) He has been on a mission to have Congress put on the national agenda the _____ about establishing a commission on aging.

9) Every autumn the stores start to _____ the big Christmas shopping season by hiring more employees and increasing advertising.

10) When he retired in 2006, there was a widespread belief that Japan was _____ for economic modernization.

11) Shops _____ the do-it-yourself craze not only by running special advisory services for novices, but by offering consumers bits and pieces which they can assemble at home.

12) It is estimated that the public _____ cuts announced this week will reduce jobs by at least 50,000, and probably over 100,000.

13) I was always able to create money by cutting the neighbors' grass, and I always saved _____ my profits.

14) She said it required a new _____ to managing resources to eradicate poverty and address conflicts in developing countries.

15) Generally speaking, we tend to be more aggressive when we're young and more _____ as we get older.

16) The question of agency is still _____ and we hope you will continue your effort to push the sale of our product at present stage.

2. Translate the following passage into Chinese.

350 million jobs are linked to marine fisheries, with 90% of fishers living in developing countries. The value of fish traded by developing countries is estimated at $ 25 billion, making it their largest single trade item. Global catch rose from 4 million tons in 1900, through 16. 7 million tons in 1950, 62 million tons in 1980 to 86. 7 million tons in 2000 but has stagnated subsequently. In 2009 marine capture production was 79 million tons. Overall catch risks decline with 75% of stocks fully exploited or depleted. Human activity has directly and markedly reduced ocean productivity; additional deficits may be due to climate change increasing ocean stratification and reducing nutrient mixing in the open seas. Global Ocean Observing System (GOOS) and LME assessments show significant warming trends from which model projections 2040—2060 forecast a steady decline in ocean productivity.

3. Translate the following passage into English.

马拉加附近海域正在测试一项颇具争议的海底采矿新技术,这项新技术让我们看到了技术的进步,证实了深海采矿确实可行,但同时也把一道难题摆在了我们面前:这样做,对吗? 这种矿石富含重要金属,尤其是电池所需的钴,所以,越来越多的人支持进行开采活动,但是对于深海采矿是否会对海底生命产生影响,目前我们还知之甚少。我们需要权衡,潜入深海,改变深海地貌以及深海生物,这样做值得吗? 像滥用土地那样开发海洋资源,这个后果我们能承担吗?

Chapter 9

Marine Laws

Unit 1

The International Law of the Sea

Formation and Development of the
International Law of the Sea

1. From the view of historical development, people's understandings of oceans and seas and their values have evolved in the following four stages. Accordingly, related rules and regulations have formed during human activities in different stages. In the first stage, oceans and seas were open to all and people could freely use marine resources. During the thousands of years from ancient times to the 15th century, people's knowledge and utilization of oceans and seas were quite limited. By the 15th century, those who had close encounters with oceans and seas were mainly coastal **inhabitants**. From their daily experiences of depending on the nearby marine environment for food and the practices of sailing, they realized that oceans and seas could provide profits of fish and salt and transportation convenience of sailing. During this historical period, people's capacities for action and the range of their activities at ocean or sea were all limited. They lived in peace with each other and also with oceans and seas which were also free.

2. In the second stage, oceans and seas became a convenient channel for exploration and **colonial plunder**. Rules and regulations about the freedom of **navigation**, the control of coastal states over coastal areas and other issues gradually took shape. From the 15th to the late 19th century, they were used as critical arteries of communication by European powers to discover and conquer the "New World". During the latter half of the 15th century, the western explorers **ushered** in the era of worldwide **explorations** and the Great Navigation Epoch. People opened up new ship routes, launched global navigation, discovered the New World, enlarged world market and started modern colonial plunder. These all promoted the development of European capitalism.

3. During this period, the most important maritime activities were explorations, voyages, fishing in coastal areas of other states. People's capacity for utilizing oceans and

seas were **tremendously** improved and their scope of activities also expanded from near-shore areas to oceans and seas around the whole world. All powers invaded and colonized many countries and regions to plunder and **accumulate** great wealth via convenient marine passages and channels. **Strategically** important sea routes became the objective over which all parties struggled. International law of the sea gradually formed and developed **considerably**. People signed many international conventions relating to the regulations of the sea warfare, navigation and marine scientific research and other marine activities.

4.   In the third stage, oceans and seas became a bone of contention in the struggle for maritime **hegemony** and marine enclosure between coastal states. Rules and regulations of the modern law of the sea had basically established. From the later 19th century to the 1970s, important marine passages and routes continued to be the target of struggle between powerful nations. There had been a wave of "marine enclosure" in which all coastal states aimed to **incorporate** nearby sea waters into their own national **sovereignty** and scope of **administration**. At the end of 19th century, Alfred Thayer Mahan put forward the concept of "sea power" which initiated a new era for mankind to know the sea. People realized that seas and oceans are important space for the survival of mankind and national security. Since WWI, people's utilization of the sea was deepened further more. In times of war, seas and oceans become important places to station troops and battlefields to fight. In peacetime, they are significant spaces used as food base, oil and gas exploitation base, tourism and recreation base, and storage base. The values of seas and oceans were explored to an even larger potential. They became the real space for the survival of mankind. They themselves became the target for states to struggle for.

5.   With more maritime rights states have claimed and more maritime activities they engaged in, the international community has organized and **compiled** relevant laws and regulations of the sea through multiple ways. The international legal system of the sea has been continuously improved. Besides the discussion of the issue of territorial waters in the drafting and passing of the Law of Territorial Water and Maritime Jurisdiction in Peacetime (draft) in 1926 and 1928 by International Law Association, the compilation of the law of the sea was mainly carried out at the League of Nations Codification Conference in 1930 in Hague by the **predecessor** of the United Nations, League of Nations and at the three United Nations Conferences on the Law of the Sea held by the UN. Related international organizations such as the United Nations contributed a lot to the integration and development of international law of the sea. For example, the 1958 Geneva Conventions on the Law of the Sea include Convention on the Territorial Sea and Contiguous Zone, Convention on the Continental Shelf, Convention of the High Seas, and Convention on Fishing and Conservation of Living Resources of the High

Seas. These conventions **regulate** people's maritime activities in precise frameworks.

6. Since WWII, the development of the law of the sea made dramatic breakthroughs. In 1947, the United Nations General Assembly decided to establish the International Law Commission for the "promotion of the progressive development of international law and its **codification**". At the first session held in 1949, the commission decided on 14 topics for the codification of international law among which the one related to the law of the sea is about high seas and territorial waters. The first session decided to set up an ad hoc committee in charge of codifying related reports. In 1954, the United Nations General Assembly required the International Law Commission to systematize the articles and clauses about high seas, territorial waters, contiguous zones, continental shelf, and conservation of marine biological resources that it had drafted. In 1956, the International Law Commission finished the final reports on the drafting the law of the sea (73 articles in total) and **submitted** them to the United Nations General Assembly. On February 21st 1957, the latter decided to hold an international conference involving states and governments to review the issue of the law of the sea.

7. From February 24th to April 27th, the United Nations held its First Conference on the Law of the Sea (UNCLOS I) at Geneva, Switzerland. Delegations from 86 nations attended the conference. Due to the wide range of tasks, the Conference set up five committees in charge of discussing and revising the articles and clauses in each part respectively. Ultimately, UNCLOS I resulted in four treaties concluded in 1958: Convention on the Territorial Sea and Contiguous Zone, Convention on the High Seas, Convention on Fishing and Conservation of Living Resources of the High Seas, and Convention on the Continental Shelf. They are called Four Geneva Conventions. Besides, the Conference also passed a Protocol of Signature on Mandatory Settlement of Disputes. Since the above-mentioned four conventions failed to resolve some important maritime issues such as territorial waters and breadth of continental shelf, the United Nations decided to convene the second Conference on the Law of the Sea (UNCLOS II).

8. From March 17th to April 27th, the United Nations held the Second Conference on the Law of the Sea (UNCLOS II) in Geneva, Switzerland. The participants were delegations from 88 nations and representatives from several specialized agencies of the United Nations and international organizations. Since it was too close to the first Conference, disagreements among related nations could not be resolved in such a short period of time. This Geneva conference did not result in any new agreements.

9. In the fourth stage, seas and oceans are the main component of the life support sys-

tem for mankind. Modern international law of the sea has developed in leaps and bounds. Since 1980s especially after the United Nations Conference on Environment and Development (UNCED) in 1992, it has been generally acknowledged in the international community that oceans and seas are an important component of the life support system for mankind and also valuable asset for sustainable development. While accumulating wealth from the sea, utilizing it to compete for wealth, depending on it for survival, people have formed a new understanding towards it, that is, they should treat it well and protect it.

10. The Third Conference on the Law of the Sea was convened in New York on December 3rd 1973. The conference lasted for 9 years until December 10th 1982 when the United Nations Convention on the Law of the Sea (UNCLOS) was opened for signature in the city of Montego Bay, Jamaica. 11 sessions and 16 meetings were convened. Delegations from 167 nations participated successively. Representatives from over 50 bodies including national liberation organizations, international organizations, and dependent territories and so on attended the conference as observers. This diplomatic conference involved the largest number of participants lasting for the longest period of time with the largest in scale. From the day of its opening for signature to the deadline on September 9th in 1984, 159 nations and bodies including China signed the United Nations Convention on the Law of the Sea (UNCLOS). The resulting convention came into force in 1994.

(1,527 words)

➤ *New Words*

inhabitant [ɪnˈhæbɪtənt]
    n. a person who inhabits a particular place 居民；居住者

Locals claim that there are enough lakes and rivers in the region for each inhabitant to have one of each.

本地人说，这里的湖泊和河流多得可以让每个居民分得一份。

colonial [kəˈləʊnɪəl]
    a. relating to countries that are colonies, or to colonialism 殖民地的

During British rule the region formed the backbone of the colonial economy and served as a cradle of Zambian nationalism.

英国统治时期，该地区是这块殖民地的经济支柱，并成为赞比亚民族独立运动的发源地。

plunder [ˈplʌndə(r)]
    n. the act of taking valuable things from a place using force 掠夺；掠夺的财物

When most of the crew was on leave from the ship, Morgan ran off with all of the plunder from Panama.

当船上大多数的船员休假时，摩根独自带着从巴拿马劫掠的战利品溜走。

navigation [ˌnævɪˈgeɪʃn]
    n. the guidance of ships or airplanes from place to place 航行；航海

It plays an increasingly important part in meteorology for cloud, precipitation, hail and thunderstorm detection as well as the navigation of aircraft and ships.

它对云层、降水、冰雹和雷暴的气象探测以及在船舰和飞机的航行中，起着日益重要的作用。

usher [ˈʌʃə(r)]
    v. to show them where they should go by going with them 引领

The developments could usher in a new age of high-speed computing in the next few years for home users frustrated with slow-running systems.

这一技术标志着在未来数年里，家庭电脑用户将迎来电脑高速发展时代，将不再因低速发展系统而受挫。

exploration [ˌekspləˈreɪʃn]
    n. a thorough examination or discussion of a subject, idea, etc. 探索

Greg pursued his dreams of space exploration all the way through college, where he majored in physics.

格雷格大学期间一直在追求他的太空探索梦想,在那里他主修物理。

tremendously [trə'mendəslɪ]	*ad.*	extremely 极大地;非常地

There has been a tremendously successful program in vaccinating children against measles in Africa, so the incidence of disease and death from that has fallen dramatically.

在非洲,在对儿童进行麻疹免疫方面有一些非常成功的项目,使麻疹发病率和死亡率大幅下降。

accumulate [ə'kjuːmjəleɪt]	*v.*	to get or gather together 累积;积聚

The pollutants disperse easily across wide geographic areas, retain their toxicity, and have a tendency to accumulate in the fatty tissues of organisms.

这种污染物很容易在大范围的地理区域内扩散,保持其毒性,并具有在生物脂肪组织内积累的倾向。

strategically [strə'tiːdʒɪklɪ]	*ad.*	with regard to strategy 战略性地;战略上

Strategically we should despise all our enemies, but tactically we should take them all seriously.

在战略上我们要藐视一切敌人,在战术上我们要重视一切敌人。

considerably [kən'sɪdərəblɪ]	*ad.*	to a great extent or degree 相当地;非常地

Sperm whales spend considerably more time at the sea surface so they too are becoming sunburned.

抹香鲸会花费相当多的时间在海面上,因此它们也会被晒伤。

hegemony [hɪ'dʒemənɪ]	*n.*	a situation in which one country, organization, or group has more power, control, or importance than others 霸权

China did not seek expansion in history, nor does it have a hegemony culture. What we have is the desire to develop ourselves and build a friendly neighbourhood.

中国没有扩张的历史,也没有称霸的文化,有的只是建设国家的需求和睦邻友好的信念。

incorporate [ɪn'kɔːpəreɪt]	*v.*	to make into a whole or make part of a whole 包含;把……合并

The agreement would allow the rebels to be incorporated into a new national police force.

该协议将允许叛军并入一支新的国家警察部队。

sovereignty ['sɒvrəntɪ]	*n.*	the power that a country has to govern itself or another country or state 主权;统治权

We maintain that the independence, sovereignty, unity and territorial integrity of each nation should be respected.

我们主张,应尊重每一个国家的独立、主权、统一和领土完整。

administration [əd͵mɪnɪˈstreɪʃn] *n.* the range of activities connected with organizing and supervising the way that an organization or institution functions 管理;行政

Generally speaking, the uprising or downfalling of an enterprise lies in its management or administration.

通常情况下,企业兴衰主要在于经营管理。

compile [kəmˈpaɪl] *v.* to produce something by collecting and putting together many pieces of information 编制;编辑

Once you compile the code, you need to deploy it in the server to access the service.

您一旦编译完了这些代码,就需要在服务器上部署它来访问服务。

predecessor [ˈpriːdəsesə(r)] *n.* one who precedes you in time (as in holding a position or office) 前任;前辈

There is some truth to this, and I certainly hope that President Obama will be far more engaged than his predecessor in tackling this agenda.

这有一定道理,我当然希望奥巴马总统在处理这个议程方面比他前任投入得多。

regulate [ˈreɡjuleɪt] *v.* to bring into conformity with rules or principles or usage 调节;规定;控制

Under such a plan, the government would regulate competition among insurance companies so that everyone gets care at lower cost.

根据这样一个计划,政府将监管保险公司之间的竞争,这样每个人都能以较低价格获得保险。

codification [͵kəʊdɪfɪˈkeɪʃn] *n.* the act of codifying 编纂;整理

The codification of the laws began in the 1840s.

这些法律的编纂始于 19 世纪 40 年代。

submit [səbˈmɪt] *v.* to hand over formally 呈递;提交

If you want your picture included at the conclusion of your article or tutorial, please submit an unretouched digital photograph.

如果您希望在文章或教程的结束部分附上您的照片,请提交一张未经修饰的数码照片。

➤ *Phrases and Expressions*

rules and regulations 规章制度
in peace with 与……和睦相处
take shape 形成;成形
a bone of contention 争论的原因
put forward 提出(观点、建议、方案、办法、理论等)
engage in 从事于;忙于
in charge of 负责;主管

➤ *Terminology*

marine enclosure 海上圈地
sea power 海权论
the articles and clauses 条文

➤ *Proper Names*

The Law of Territorial Water and Maritime Jurisdiction in Peacetime (draft) 和平时期海上管辖
　　权法公约(草案)
International Law Association 国际法学会
League of Nations Codification Conference 国际联盟编纂会议
three United Nations Conferences on the Law of the Sea 联合国三次海洋法会议
1958 Geneva Conventions on the Law of the Sea 1958 年日内瓦海洋法公约
Convention on the High Seas 公海公约
Convention on the Territorial Sea and Contiguous Zone 领海及毗连区公约
Convention on the Continental Shelf 大陆架公约
Convention on Fishing and Conservation of Living Resources of the High Seas 公海捕鱼和生物资
　　源生物资源养护公约
United Nations General Assembly 联合国大会
International Law Commission 国际法委员会
First Conference on the Law of the Sea 第一次海洋法会议
United Nations Conference on Environment and Development (UNCED) 联合国环境与发展会议

➤ *Translation*

1. From their daily experiences... transportation convenience of sailing. (Para. 1)
　　人们通过靠海吃海和就近航海的实践,对海洋有了"鱼盐之利和舟楫之便"的认识。
　　【在翻译的过程中,完全按照原文的句式和字典的释义来进行翻译是行不通的。汉英两种语言在文化背景和语法结构上存在着很大的差异,因此,我们必须根据具体的上下文进行灵活处理。此句译文的灵活性不仅体现在四字格的妙用,还体现在词性转换和语序调整上。】

2. From the 15th to the late 19th century... conquer the "New World". (Para. 2)

15 世纪至 19 世纪末,海洋成为欧洲强国发现和征服"新大陆"的重要通道。

【同一个词语在不同的上下文中,可能含有不同的意义。这就需要我们依靠具体的上下文来判断某一个词语的意义。"arteries"一词有"动脉;干道;主流"这几个释义,根据上下文选取"干道"这个释义,引申为通道。另外此句译文还使用了转态译法(被动变主动)和换序译法(语序调整)。】

3. People's capacity for utilizing oceans... around the whole world. (Para. 3)

人类对海洋的利用能力极大地增强,利用海洋的范围也从近岸扩展到全球海洋。

4. There had been a wave of "marine enclosure"... scope of administration. (Para. 4)

沿海国掀起了将邻近海域划入国家主权和管辖范围的"海上圈地"热潮。

5. At the end of 19th century... a new era for mankind to know the sea. (Para. 4)

19 世纪后期,阿尔弗雷德·塞耶·马汉提出的"海权论"开创了人类认识海洋的新时代。

6. With more maritime rights... regulations of the sea through multiple ways. (Para. 5)

随着各国在海洋上的权利要求和所从事的活动越来越多,国际社会先后通过多种方式,对相关海洋规则进行整理编纂。

7. Besides the discussion of the issue of territorial waters... held by the UN. (Para. 5)

除了 1926 年、1928 年国际法学会讨论通过领海问题并草拟《和平时期海上管辖权法公约(草案)》之外,关于海洋法的编纂活动主要是通过联合国的前身国际联盟于 1930 年召开的海牙国际法编纂会议以及联合国三次海洋法会议进行的。

8. At the first session held in 1949... high seas and territorial waters. (Para. 6)

1949 年国际法委员会在其第一届会议上,选定了 14 个编纂题目,其中与海洋法有关的是公海和领海制度。

9. Since the above-mentioned four conventions... the Sea (UNCLOS II). (Para. 7)

由于上述 4 个公约未能解决领海和大陆架宽度等一些重要海洋问题,联合国决定召开第二次海洋法会议。

10. Since 1980s especially... valuable asset for sustainable development. (Para. 9)

自 20 世纪 80 年代以来,特别是 1992 年世界环境与发展大会之后,国际社会普遍认识到,海洋是人类生命支持系统的重要组成部分,也是可持续发展的宝贵财富。

 **Exercises**

1. Fill in the blanks with the proper given words, and then translate the sentences into Chinese.

accordingly	encounter	capacity	era
route	scope	session	initiate
draft	resolve	territorial	treaty
jurisdiction	community	convene	asset

1) We've learnt to _____ the dialogue with visitors, involve them into discussions and gain their trust by addressing their needs and speaking with them honestly and directly.

2) We have a different background and a different history. _____, we have the right to different futures.

3) Under the _____, countries with nuclear weapons agree to move toward disarmament, while countries without nuclear weapons agree not to acquire them, and all have the right to peaceful nuclear energy.

4) Any foreign company's oil and gas exploration activity in the waters under China's _____ without China's permission is illegal and invalid.

5) We live in an _____ in which information, goods and capital speed around the globe, every hour of every day.

6) If we separate a girl and isolate her from her _____, what will her life be like?

7) After two late night _____, the Security Council has failed to reach agreement before Friday.

8) We maintain that the independence, sovereignty, unity and _____ integrity of each nation should be respected.

9) Every day of our lives we _____ major and minor stresses of one kind or another.

10) He began cracking open big blue tins of butter cookies and feeding the dogs on his _____.

11) So you're saying this will strengthen the American public's _____ to go to war if necessary?

12) In September, the General Assembly will _____ a High-level Meeting on the prevention and control of non-communicable diseases.

13) This requires IT to no longer be viewed as a utility but rather as an integral and vital _____ to the company.

14) Huntington asked me to comment on a _____ of the essay while I was his graduate student.

15) In 1860 slaves as an asset were worth more than all of America's manufacturing, all of the railroads and all of the productive _____ of the United States put together.

16) Additionally, disaster recovery is a complex topic beyond the _____ of this discussion, and should be addressed by experts.

2. Translate the following passage into Chinese.

The Third Conference on the Law of the Sea was convened in New York on December 3rd 1973. The conference lasted for 9 years until December 10th 1982 when the United Nations Convention on the Law of the Sea (UNCLOS) was opened for signature in the city of Montego Bay, Jamaica. 11 sessions and 16 meetings were convened. Delegations from 167 nations participated successively. Representatives from over 50 bodies including national liberation organizations, international organizations, and dependent territories and so on attended the conference as observers. This diplomatic conference involved the largest number of participants lasting for the longest period of time with the largest in scale. From the day of its opening for signature to the deadline on September 9th in 1984, 159 nations and bodies including China signed the United Nations Convention on the Law of the Sea (UNCLOS).

3. Translate the following passage into English.

海洋面积近3.6亿平方千米,约占地球表面总面积的71%。在世界海洋中,有一部分属于沿海国家的管辖海域,面积约为1.09亿平方千米,约占世界海洋面积的30%;其余部分的海域属于沿海国家管辖之外的,面积约有2.5亿平方千米。海洋的中心主体部分、面积广阔的深水水域被称为洋,约占海洋总面积的89%,深度一般在3 000米以上。地球上有四大洋:太平洋、大西洋、印度洋和北冰洋。太平洋约占大洋总面积的二分之一;大西洋次之,面积约为太平洋的一半;印度洋第三,面积约为7 900万平方千米。

Unit 2

The United Nations Convention on the Law of the Sea

The Relations between the International Law, the International Law of the Sea, and the Convention

1. Oceans and seas have already become a hot issue in international affairs and the focus of attention of coastal states. With the development of marine scientific research and innovation of technical means, people are deepening their understanding of the strategic position and values of oceans and seas. They have developed more powerful capacities to explore and utilize the resources of oceans and seas. Various kinds of maritime disputes are gradually growing. People encounter increasing pressure to protect marine ecological environment. It has been widely acknowledged among the international community that it is of **paramount** significance to establish and improve international marine order for the peaceful utilization and protections of oceans and seas. Therefore, the international law of the sea has become the most rapidly developed department law in modern international law. The United Nations Convention on the Law of the Sea (UNCLOS) passed in 1982 (**hereinafter** referred to as the 1982 Convention or the Convention) at the third Conference on the Law of the Sea is a concentrated reflection of international law of the sea.

2. As for the relations between international law, international law of the sea, and 1982 United Nations Convention on the Law of the Sea (UNCLOS), it should be pointed out that the international law is a general term. It is at the **spire** of a tower-like legal system. At the lower part, there are public international laws, private international laws (conflict of laws), international trade laws, and so on. Public international law can be subdivided into international law of the sea, international air law, space law, international criminal law, law of war and other department laws.

3. International law was originated from the west and based on the western philosophy of law. There are several important legal ideas in the western philosophy of law worth mentioning and understanding. The first is the principle of "*Ei incumbit probation quidicit, non qui negat*" (Latin, "the burden of proof is on he who declares, not on he who denies") and of "**presumption** of innocence". In plainer words, when accusing the **defendant** of some crime, the burden of proof is on the **prosecution**. The accuser should provide convincing evidences and specific articles of law to prove the illegal actions of the defendant. On the other hand, if the accuser fails to prove the crime or fault of the defendant or against which legal rules or regulations the defendant acts, the rights that the defendant claims and his or her actions are legally right naturally. The second is "*Nullum crimen, nulla poena sine praevia lege poenali*" (Latin, "[There exists] no crime [and] no punishment without a preexisting penal law [**appertaining**]"). Actions not expressly **prohibited** in legal provisions are naturally legally **feasible**.

4. As a major component of the international law, the international law of the sea can be subdivided into various kinds of international treaties and conventions concerning marine affairs, such as the above-mentioned four Geneva Conventions in 1958, 1982 Convention and other international treaties and rules and regulations of **customary** laws of the sea.

5. Generally speaking, the relation between the international law, the international law of the sea, and the 1982 Convention are similar to that between higher laws and lower laws such as that between the Constitution of China and other department laws or between parent laws and department laws in China's legal system. Some principles of international law have also been taken as regulations of international law of the sea. Therefore, they have also been written into the 1982 Convention. Any **interpretation** and application of all provisions of the 1982 Convention must not violate the general theories, principles and regulations of international law. To be specific, there should be no violation of the Vienna Convention on the Law of Treaties (VCLT) or rules and regulations concerning the interpretation of international treaties.

6. Therefore, any international convention can only address issues of several aspects in a certain field. There is no cure-all. The 1982 Convention is a major component of the international law of the sea which in turn is a department law and also an important component of the international law. In other words, the United Nations Convention on the Law of the Sea (UNCLOS) is a component of international law of the sea, which in turn is **subordinate** to international law. However, it is impossible to use the 1982 Convention to replace the international law of the sea or the international law. For instance, the correct way should be "to settle disputes of the South China Sea with in-

ternational law" or "to settle disputes of the South China Sea with international law including the United Nations Convention on the Law of the Sea (UNCLOS)". The expression of "to settle disputes of the South China Sea with the United Nations Convention on the Law of the Sea (UNCLOS)" is wrong and devoid of common sense of international law.

7. Above all, the rules and regulations people abide by when engaged in various kinds of marine activities, the **demarcation** of sea areas under and beyond national jurisdiction, the rights states can enjoy and obligations they should fulfill in different sea waters and other issues are what concern the United Nations Convention on the Law of the Sea (UNCLOS). However, as an international convention, the Convention "recognizes the **desirability** of establishing through this Convention, with due regard for the sovereignty of all states, a legal order for the seas and oceans which will **facilitate** international communication, and will promote the peaceful uses of the seas and oceans, the equitable and efficient utilization of their resources, the conservation of their living resources, and the study, protection and preservation of the marine environment". This indicates that the Convention only stipulates relevant rules and regulations concerning the above-mentioned issues but does not attempt to solve all marine issues.

(1,002 words)

➢ *New Words*

paramount [ˈpærəmaʊnt]

a. having superior power and influence 最重要的；至高无上的

These are basic industries in which closeness to raw materials and markets for their bulky products is paramount.

对于这些基础产业来说，保持自身庞大的产品与原材料和市场的紧密联系是至关重要的。

hereinafter [ˌhɪərɪnˈɑːftə(r)]

ad. in a subsequent part of this document or statement or matter, etc. 在下文中

The recently closed Spring Meeting of International Monetary Fund and World Bank (hereinafter abbreviated to IMF and WB) will leave an important stroke on the history of international economic order.

刚刚闭幕的国际货币基金组织和世界银行（以下简称基金组织和世行）春季会议将在国际经济秩序演进的历史中留下重要一笔。

spire [ˈspaɪə(r)]

n. a tall tower that forms the superstructure of a building and that tapers to a point at the top 尖顶；尖塔

In the distance, the golden spire of the Leifeng Pagoda, a stunning 10th-century landmark rebuilt only eight years ago, emerges out of the greenery.

在远处，雷峰塔的尖顶显现在绿色的丛林之上。这座壮观的地标建筑始建于 10 世纪，刚刚在 8 年前重建完毕。

presumption [prɪˈzʌmpʃn]

n. something that is accepted as true but is not certain to be true 假定；推测

The presumption is that parents who can afford fee-paying schools can buy a pure-wool suit as well.

这背后的假设是，有能力负担私立学校学费的家长，也能负担纯羊毛西装。

defendant [dɪˈfendənt]

n. a person or institution against whom an action is brought in a court of law; the person being sued or accused 被告

But for capital offences, a full commission composed of judge and jury must still be convened to convict and sentence the defendant.

但是对于死刑判决，必须组织一个由法官和陪审团组成的完整的法庭，然后才能对被告进行宣判。

prosecution [ˌprɒsɪˈkjuːʃn]

n. the action of charging someone with a crime and putting them on trial 起诉

Yesterday the head of government called for the prosecution of those responsible for the deaths.

政府首脑昨日要求对那些造成死亡的责任人进行起诉。

appertain [ˌæpəˈteɪn]　　*v.* to be a part or attribute of 属于;和……有关

Clearly, however, these matters appertain as much to private economy as to the economy of whole nations.

但是很明显,这些问题关系到整个国家的经济,也在同样程度上关系到私营经济。

prohibit [prəˈhɪbɪt]　　*v.* to stop something from being done 禁止

a law prohibiting the sale of alcohol 禁止售酒的法令

feasible [ˈfiːzəbl]　　*a.* capable of being done with means at hand and circumstances as they are 可行的

WHO fully agrees with FAO and OIE that control of avian influenza infection in wild bird populations is not feasible and should not be attempted.

世卫组织完全同意粮农组织和国际兽疫局的观点,即在野鸟种群中控制禽流感感染既不可行也不应尝试。

customary [ˈkʌstəmərɪ]　　*a.* in accordance with convention or custom 习惯的;通常的

Cafes and coffee houses are an Austrian tradition, and it is customary to take an afternoon break for a strong cup of coffee.

咖啡吧和咖啡馆是奥地利人的一项传统,他们习惯在下午休息的时候,去喝上一杯浓咖啡。

interpretation [ɪnˌtɜːprəˈteɪʃn]　　*n.* an opinion about what it means 解释

Yet if that is the right interpretation, it leaves the question of what evolved into the trilobites and their kind as mysterious as ever.

然而即使这样的解释是正确的,三叶虫和它们的同类由何而来的问题仍然神秘晦涩,一如往昔。

subordinate [səˈbɔːdɪnət]　　*a.* lower in rank or importance 从属的;次要的

Local people's governments at various levels throughout the country are state administrative organs under the unified leadership of the state Council and are subordinate to it.

全国各级地方人民政府都是国务院统一领导的国家行政机关,隶属于国务院。

demarcation [ˌdiːmɑːˈkeɪʃn]　　*n.* the establishment of boundaries or limits separating two areas, groups, or things 划界

It was hard to draw clear lines of demarcation between work and leisure.

工作和闲暇之间很难划出明确的界限。

desirability [dɪˌzaɪərəˈbɪlətɪ] *n.* the quality of being worthy of desiring 可取性

The unique nature of gold and precious metals provides its desirability in this Fed operation.

黄金和贵金属所具有的独特性质为美联储的这一动作提供了可取性。

facilitate [fəˈsɪlɪteɪt] *v.* to increase the likelihood of (a response) 促进;使容易

All of these agencies and people either complement or facilitate our work, or express the health needs of populations.

所有这些机构和人士要么补充或便利我们的工作,要么表达人民的卫生需求。

➤ *Phrases and Expressions*

be originated from 起源于
accuse... of... 因某事控告某人
legally right 合法的
legal provisions 法律规定
in turn 反过来;依次
devoid of 没有;缺乏
common sense 常识
abide by 遵守

➤ *Terminology*

department law 部门法
public international laws 国际公法
private international laws 国际私法
international trade laws 国际贸易法
international law of the sea 国际海洋法
international air law 国际航空法
space law 外空法
international criminal law 国际刑法
law of war 战争法
penal law 刑法
customary laws of the sea 习惯海洋法

parent laws 母法

➤ *Proper Names*

the Constitution of China 中国宪法

➤ *Translation*

1. With the development of marine... position and values of oceans and seas. (Para. 1)
   随着海洋科学研究和技术手段的发展,人类对海洋战略地位及其价值的认识不断深化。

2. The United Nations Convention... of international law of the sea. (Para. 1)
   1982 年第三次联合国海洋法大会通过的《联合国海洋法公约》(以下简称《公约,1982》或《公约》)成为国际海洋法的集中体现。

3. The first is the principle of... and of "presumption of innocence". (Para. 3)
   一是"谁指控,谁举证"和"无罪推定原则"。
   【翻译过程中,把原文中需要而译文中不需要的单词、词组等在翻译过程中加以省略。这种省词译法一般是出于译文语法和习惯表达法的需要。此句引号中的拉丁文和引号中的英文含义重复,所以翻译时先省略引号中的拉丁文。其次如果保留短语"not on he who denies",译文就显得拖泥带水,甚至会出现画蛇添足的结果。】

4. On the other hand, if the accuser... actions are legally right naturally. (Para. 3)
   换句话说,只要原告无法证明被告到底错在哪儿、违背了哪一条法律规则或原则,那么,被告所主张的权利和所采取的行动就自然都是合法的。
   【一般说来,英美国家的人们比较喜欢用名词和介词,而汉语里动词用得比较多一些。因此,英译汉时,我们经常要把英语中的名词和介词转换成汉语中的动词。所以,翻译此句时,把名词短语"crime or fault"翻译成动词"(犯)错",把介词"against"翻译成"违背"。】

5. The second is "Nullum crimen... preexisting penal law [appertaining]"). (Para. 3)
   二是"法无明文禁止即可为"。

6. As a major component... regulations of customary laws of the sea. (Para. 4)
   国际海洋法是国际法的一个重要组成部分。在国际海洋法下面,又可细分出各类涉及海洋事务的国际条约,例如,1958 年日内瓦四公约和 1982 年公约等国际条约,以及习惯海洋法规则等内容。

7. The 1982 Convention is a major component... international law. (Para. 6)
   《公约,1982》是国际海洋法的一个重要组成部分,而国际海洋法是国际法中的一个部门法,也是国际法的一个重要组成部分。

8. For instance, the correct way... on the Law of the Sea (UNCLOS)". (Para. 6)
   例如,正确的说法是"应该用国际法解决南海争端",或者是"应该用包括《联合国海洋法公约》在内的国际法解决南海争端"。

9. Above all, the rules and regulations... on the Law of the Sea (UNCLOS). (Para. 7)

综上所述,人们在海洋从事各种活动应遵守哪些制度和规则、如何划分国家管辖海域和管辖范围外海域、各国在不同海域里分别拥有什么权利和义务等问题,是《联合国海洋法公约》所规定的基本内容。

10. However, as an international convention... the marine environment". (Para. 7)

但是,作为一个国际公约,《公约》也只是为了"在妥为顾及所有国家主权的情形下,为海洋建立一种法律秩序,以便利国际交流,促进海洋和平利用,促进海洋资源公平而有效地开发,促进海洋生物资源的养护以及海洋环境的研究、保护和保全"。

 Exercises

1. *Fill in the blanks with the proper given words, and then translate the sentences into Chinese.*

ecological	convention	reflection	general
criminal	worth	evidence	deny
accuse	claim	fulfill	address
engage	component	conservation	attempt

1) Youth violence greatly increases the costs of health, welfare and _____ justice services, reduces productivity, decreases the value of property, and generally undermines the fabric of society.

2) Mastery of this line at work is key to proving your _____ to those around you.

3) Over 2,000 assemblers and _____ suppliers from 14 different countries are showcasing their products this year.

4) We did it so we could have more time to follow our passions, to realize our dreams, and to _____ our deepest desires and longings.

5) Communication is something we all _____ in on a daily basis but due to the pace of our lives, conversations become just formalities.

6) More than half of all forests are used for wood and non-wood production in combination with other functions such as soil and water protection, biodiversity _____ and recreation.

7) The supersession of the old by the new is a _____, eternal and inviolable law of the universe.

8) They are open, flexible systems that can be moved or modified with changes in a society's needs or in the environment, and in that sense they are _____, systems-based, and socially responsible.

9) We will work with them, as we will work with you, to _____ our common concerns.

10) He has gathered _____ from mice to support this idea, but whether it is the case in people has yet to be tested.

11) Villagers _____ the company of destroying their fields to expand the banana plantations after getting land leases from the government.

12) The only time that we _____ to do something like that was in the city of Philadelphia.

13) We tend to regard hair only as a _____ of our personal style, but as a living, growing part of our body, it's also tied in to our physiology.

14) Overly proud people refuse to admit their mistakes and sometimes even _____ the fact that they make them because they feel that it would impugn their image.

15) John had taken out insurance but when he tried to _____, the insurance company refused to pay.

16) Some 150 states and regional integration organizations have signed the _____, and 100 have ratified it.

2. Translate the following passage into Chinese.

Generally speaking, the relation between the international law, the international law of the sea, and the 1982 Convention are similar to that between higher laws and lower laws such as that between the Constitution of China and other department laws or between parent laws and department laws in China's legal system. Some principles of international law have also been taken as regulations of international law of the sea. Therefore, they have also been written into the 1982 Convention. Any interpretation and application of all provisions of the 1982 Convention must not violate the general theories, principles and regulations of international law. To be specific, there should be no violation of the Vienna Convention on the Law of Treaties (VCLT) or rules and regulations concerning the interpretation of international treaties.

3. Translate the following passage into English.

中国是世界上最大的发展中沿海国,濒临太平洋西侧 4 个边缘海——渤海、黄海、东海和南海。中国是包括《公约,1982》在内的许多国际海洋条约的缔约国。中国积极参与《公约》的制定过程以及后续相关规章的发展进程。全国人民代表大会常务委员会于 1996 年 5 月 15 日通过了《关于批准〈联合国海洋法公约〉的决定》。依据《公约,1982》,中国与其他缔约国在全球海洋和周边海洋都享有相同的权利,并履行同等的义务。同时,中国也与其他国家一样,在海洋里还享有除了《公约》之外的其他国际法所赋予的一切权利,并承担相关的义务。

Unit 3

The Law of the Sea and International Organizations

Development of the Law of the Sea
through International Organizations

1. The role of international organizations is increasingly important in international law, and the same applies to the law of the sea. Notably, several international organizations, including the UN "family", make an important contribution to the development of the law of the sea. In this regard, the best example may be the International Maritime Organization (IMO). The IMO has a wide jurisdiction relating to the safety of navigation as well as the protection of the marine environment. To date, many instruments have been adopted under the auspices of the IMO. Those instruments have become more important after the entry into force of the LOSC, since the practice of the States Parties to the Convention shall be conformity to the international standards created through the IMO by virtue of "rules of reference". According to the "rules of reference", relevant provisions of the LOSC must be implemented in accordance with rules adopted under the auspices of the IMO, to the extent that these rules are "applicable" or "generally accepted". In so doing, IMO instruments can further **elaborate** provisions of the LOSC.

2. Another important organization in the field of law of the sea is the Food and Agriculture Organization of the United Nations (FAO). The FAO is the only organization of the UN system that has a global fisheries body, the Committee on Fisheries. The FAO thus has a **prime** role in the conservation and management of fisheries, including review of world fisheries and assistance to developing countries. At the same time, the FAO serves as the forum for discussion and negotiation of international **instruments** in this field. The instruments adopted under the auspices of the FAO may affect interpretations and implementation of the LOSC.

3. The 1995 FAO Code of Conduct on Responsible Fishing is a case in point. The Code of Conduct is global in scope, and is directed towards members and non-members of the FAO, fishing entities, subregional, regional and global organizations, whether governmental or non-governmental, and all persons concerned with the conservation of fishery resources and management and development of fisheries pursuant to Article 1. 2. Whilst the Code of Conduct is a voluntary instrument relating to fisheries, certain parts of it are based on relevant rules of international law, including those reflected in the LOSC. The Code of Conduct is to be interpreted and applied in conformity with the relevant rules of international law, as reflected in the LOSC. The Code of Conduct is also to be interpreted and applied in conformity with the 1995 UN Fish Stocks Agreement under Article 3. To this extent, in part the Code of Conduct may interpret and **amplify** relevant provisions of the LOSC as well as the 1995 Fish Stocks Agreement.

4. Next, the role of the UN General Assembly in the development of the law of the sea must be mentioned. In light of its universal membership, the UN General Assembly can provide an international forum for discussion and negotiation on the law of the sea, including the LOSC. After the entry into force of the LOSC in 1994, the General Assembly decided "to undertake an annual review and evaluation of the implementation of the Convention [LOSC] and other developments relating to ocean affairs and the law of the sea". In relation to this, the General Assembly has requested the UN Secretary-General to prepare annually a **comprehensive** report on developments relating to the law of the sea.

5. Furthermore, the United Nations Open-ended Informal **Consultative** Process (ICP) on Oceans and the Law of the Sea was established by the General Assembly Resolution of 24 November 1999. The ICP has met every year since 2002 and has established itself as a useful forum for discussions on marine affairs. The ICP is open to observers from relevant international institutions and non-governmental organizations (NGOs). At the process, independent experts are invited to participate in discussion panels which are designed to examine relevant topics. The work of the ICP is important to facilitate the annual review of developments in ocean affairs by the General Assembly. More generally, it is widely acknowledged that the UN General Assembly makes important contributions to the making of customary international law. Considering that rules of customary law governing the oceans are a matter of interest for all States beyond the circle of the Contracting Parties to the LOSC, the role of the General Assembly in customary law-making in this field will not lose its importance. Since all States have in theory an equal voice and an equal vote in the General Assembly, it can also contribute to enhanced **legitimacy** and **democratization** in the making of customary law.

6. Finally, some mention must be made of the role of NGOs in the development of the law of the sea. NGOs perform diverse functions in international relations, such as: raising public awareness, taking political **initiatives**, initiating conferences, drafting a treaty or non-binding instruments, developing and **clarifying** rules and standards, presentation of information, deployment for the implementation of treaties, **verification** and monitoring. NGOs are engaged in the formation of law and policy concerning marine affairs, although the scope of their involvement varies considerably. In this regard, it is of particular interest to note that some treaties allow NGOs to participate in meetings of the Parties as observers.

7. An **illustrative** example is provided by the 1992 Convention for the Protection of the Marine Environment in the North-East Atlantic (the OSPAR Convention). Under Article 11(1)(b) of the Convention, the OSPAR Commission may, by **unanimous** vote of the Contracting Parties, decide to admit as an observer "any international governmental or any nongovernmental organization the activities of which are related to the Convention". Such observers are allowed to participate in meetings of the OSPAR Commission but without the right to vote and to present to the Commission any information or reports relevant to the objectives of the Convention by virtue of Article 11(2). To this extent, NGOs can influence the work of the Commission concerning marine environmental protection.

(1,016 words)

➣ *New Words*

elaborate [ɪˈlæb(ə)rət]

    v. to add details, as to an account or idea 详细阐述

    The specific acquisition case you mentioned is different from the so-called trade and investment protectionism. Officials with the Ministry of Commerce have made elaborate explanation on this.

    这个具体的并购案例不同于所谓的贸易和投资保护主义,中国商务部有关官员已就此做出了非常详尽的解释。

prime [praɪm]

    a. first in rank or degree 主要的

    Portion distortion is one of the prime causes of America's obesity epidemic, yet most of them don't realize that they overeat.

    暴饮暴食是美国肥胖人口增长的主要原因之一,但是他们当中的大部分人并没有意识到他们吃过了头。

instrument [ˈɪnstrəmənt]

    n. a document that states some contractual relationship or grants some right 文件

    The primary instrument for protecting and fulfilling these rights is the United Nations Convention on the rights of the Child (CRC).

    保护和满足这些权利的主要文件就是联合国儿童权利公约。

amplify [ˈæmplɪfaɪ]

    v. to increase in size, volume or significance 扩大;增强

    A careful reader wants to argue with the author, or amplify a point, or jot down an insight inspired by something freshly read.

    一个认真的读者会想要和作者争论,或者放大一个观点,又或者赶紧把由阅读引发的新鲜感想记录下来。

comprehensive [ˌkɒmprɪˈhensɪv]

    a. including all or everything 全面的;综合的

    People want a more comprehensive answer than that, but in my experience, if you learn to do this, the rest will follow.

    人们希望得到更全面的答案,但是我的经历告诉我:如果你学会做好一件事,其他也一通百通。

consultative [kənˈsʌltətɪv]

    a. giving advice 顾问的;咨询的

The work done this week, supported by the seven consultative groups, is a solid step forward.

过去的一周里在七个顾问小组的支持下完成的工作,是向前迈出的坚实的一步。

legitimacy [lɪˈdʒɪtɪməsɪ]　　*n.* lawfulness by virtue of being authorized or in accordance with law 合法;合理

This legitimacy depends on a social contract that treats the population as citizens rather than subjects, and has as its primary goal the economic and social advancement of society.

这一合法性依赖于这样一种社会契约,即将老百姓都当成公民而非臣民,并将整个社会的经济与社会发展作为其首要目标。

democratization [dɪˌmɒkrətaɪˈzeɪʃn]　　*n.* the action of making something democratic 民主化

With the development of the economical globalization, the democratization of international relation becomes a hot problem that academic circles pay attention to gradually.

随着经济全球化的日益发展,国际关系的民主化逐渐成为学术界关注的一个热点问题。

initiative [ɪˈnɪʃətɪv]　　*n.* an important act or statement that is intended to solve a problem 法案;倡议

Local initiatives to help young people have been inadequate.

当地对年轻人的扶助法案一直不够完善。

clarify [ˈklærəfaɪ]　　*v.* to make clear and (more) comprehensible 澄清;阐明

Business analysts, for example, need not participate unless they are required to clarify something in the use case.

举例来说,业务分析师不需要参加,除非需要他们阐明用例中的某些内容。

verification [ˌverɪfɪˈkeɪʃn]　　*n.* additional proof that something that was believed 确认,查证;核实

We had to weigh these factors and select an appropriate strategy for verification and validation.

我们必须权衡这些因素,并选择一个合适的验证和确认策略。

illustrative [ˈɪləstrətɪv]　　*a.* clarifying by use of examples 说明的;作例证的;解说的

Each article will dive deep into a particular illustrative aspect of one or both of these concepts.

每一篇文章都将深入探讨其中一个或两个概念的一个特定说明性方面。

unanimous [juˈnænɪməs]　　　　　*a.* in complete agreement 一致同意的

Despite nearly unanimous acknowledgment among scientists that cancer is winning the war, optimism abounds.

尽管科学家们几乎一致承认癌症正在赢得这场战役,但乐观情绪依然高涨。

➤ *Phrases and Expressions*

rules of reference 推理规则
by virtue of 由于;凭借
in part 部分地;在某种程度上
in conformity with 符合;遵照
a case in point 恰当的例子
participate in 参加;分享

➤ *Proper Names*

International Maritime Organization（IMO）国际海事组织
Food and Agriculture Organization of the United Nations（FAO）联合国粮食及农业组织（粮农组织）
the United Nations Open-ended Informal Consultative Process on Oceans and the Law of the Sea 联合国海洋和海洋法不限成员名额非正式协商进程
NGO（non-governmental organization）非政府组织
the OSPAR Convention 保护东北大西洋海洋环境公约

➤ *Translation*

1. Notably, several international organizations... of the law of the sea.（Para. 1）
   值得注意的是,包括联合国"大家庭"在内的几个国际组织为海洋法的发展做出了重要贡献。

2. Those instruments have become more... by virtue of "rules of reference".（Para. 1）
   《公约》生效后,《公约》缔约国的做法就应符合海事组织根据"推理规则"制定的国际标准,因而这些文书变得更为重要。
   【线性思维为英语的思维特征之一,这种思维方式决定了英语句子是主干凸显的结构,即结论、新观点、新信息等句子最重要的部分往往要置于句子的主干部分,但汉语的思维方式是螺旋式思维,汉语句子往往是尾部凸显的结构,即句子最重要的部分往往要置于句子的尾部。因此进行英汉互译时,需要进行句子结构调整,如汉译本句时,最重要的信息"这些文书变得更为重要"要置于句尾。】

3. According to the "rules of reference"... or "generally accepted". (Para. 1)

根据"推理规则",只要国际海事组织主持通过的规则适用或得到普遍接受,《海洋法公约》的有关规定必须按照这些规则执行。

【英汉两种语言中都有一词多义和一词多类的现象,而辞典汉语释义又未必与英文意思完全对等,这就为准确、贴切地译词带来了困难。选择词义时,应根据上下文确定词义,表达词义时要考虑词与词的搭配。"to the extent"在词典中意为"到……的程度",但是根据本句的上下文和逻辑关系,翻译成"到……的程度"会令人费解,所以理解为"在……的范围内",翻译成"只要……"。】

4. Whilst the Code of Conduct... including those reflected in the LOSC. (Para. 3)

尽管《行为守则》是与渔业有关的自愿文书,但其中某些部分基于有关的国际法规则,包括《海洋法公约》所反映的规则。

5. The Code of Conduct... UN Fish Stocks Agreement under Article 3. (Para. 3)

《行为守则》也应按照 1995 年《联合国鱼类种群协定》第 3 条的规定进行解释和适用。

6. In light of its universal membership... law of the sea, including the LOSC. (Para. 4)

鉴于其成员的普遍性,联合国大会可以提供一个国际论坛,讨论和协商包括《海洋法公约》在内的海洋法。

7. Considering that rules of customary law... will not lose its importance. (Para. 5)

考虑到管理海洋的习惯法规则是《海洋法公约》缔约国以外所有国家都关心的问题,大会在这一领域中制定习惯法的作用将不会失去其重要性。

8. Since all States have in theory... in the making of customary law. (Para. 5)

由于理论上所有国家在大会中都有平等的发言权和平等的投票权,因此在制定习惯法方面,它也有助于加强合法性和民主化。

9. NGOs perform diverse functions in... verification and monitoring. (Para. 6)

非政府组织在国际关系中发挥多种功能,例如:提高公众意识、采取政治举措、召开会议、起草条约或非约束性文书、制定和澄清规则和标准、提供信息、部署条约的执行、核查和监测。

10. Such observers are allowed to participate... by virtue of Article 11(2). (Para. 7)

允许这些观察员参加《保护东北大西洋海洋环境公约》委员会的会议,但无表决权,也无权根据第 11 条第 2 款向委员会提交与《公约》目标有关的任何信息或报告。

 **Exercises**

1. Fill in the blanks with the proper given words, and then translate the sentences into Chinese.

entry	conformity	virtue	reference
fisheries	review	voluntary	annual
universal	request	institution	panel
contribution	vote	evaluation	adopt

1) Fiji cherishes friendship with China and expects to cooperate with China in _____, education, public health and tourism fields.

2) The Chinese chemist Tu Youyou is honoured with the Noble Prize in Physiology or Medicine for her _____ in treating malaria.

3) The _____ Product Code symbol, also known as the "bar code", is printed on products for sale and contains information that a computer can read.

4) A review _____ concluded that there were no exceptional circumstances that would warrant a lesser penalty for him.

5) The evacuation is being organized right now at the _____ of the United Nations Secretary General.

6) The inspection reveals that both the quantity and quality of the wheat delivered are not in _____ with those stipulated in the contract, though the packing is all in good condition.

7) While wealth is not synonymous with _____, a correlation exists between adding value and garnering success.

8) Several companies are offering new, _____-level models in hopes of attracting more buyers.

9) At that time, the residents of Leningrad _____ to restore the city's original name of St. Petersburg.

10) One of our contributions as a global _____ is to ensure that the world understands that and takes action accordingly.

11) His research can serve as a valuable _____ for the UN to control the spread of EBOLA, a deadly virus.

12) In recent years, young upper middle-class people in London, have begun to _____ some regional accents, in order to hide their class origins.

13) _____ work was particularly important in view of the fact that women were often forced to give up paid work on marriage.

14）Retail stores usually count on the Christmas season to make up to half of their _____ prof-its.

15）It describes only how children grow in a particular region and time, but does not provide a sound basis for _____ against international standards and norms.

16）Another study, published in the _____ of *Economics and Statistics*, reports that the effort put forth by parents is extremely important.

2. Translate the following passage into Chinese.

Another important organization in the field of law of the sea is the Food and Agriculture Organization of the United Nations (FAO). The FAO is the only organization of the UN system that has a global fisheries body, the Committee on Fisheries. The FAO thus has a prime role in the conservation and management of fisheries, including review of world fisheries and assistance to developing countries. At the same time, the FAO serves as the forum for discussion and negotiation of international instruments in this field. The instruments adopted under the auspices of the FAO may affect interpretations and implementation of the LOSC. The 1995 FAO Code of Conduct on Responsible Fishing is a case in point.

3. Translate the following passage into English.

《公约》主要设立了领海及毗连区、用于国际航行的海峡、群岛国、专属经济区、大陆架、公海、岛屿、闭海或半闭海、内陆国出入海洋的权利和过境自由、国际海底区域、海洋环境的保护和保全、海洋科学研究、海洋技术的发展和转让、争端的解决等法律制度。《公约》极大地扩展了沿海国的管辖范围，推动了国际海洋新秩序的建立。《公约》是现代海洋法的主要渊源和权威文件，被誉为"海洋宪章"，在国际海洋法的发展史上具有重要意义。《公约》还设立了三个专门的国际机构，即国际海底管理局、大陆架界限委员会和国际海洋法法庭。

Chapter 10

Marine Culture

Unit 1

Marine History

New Pathways in Maritime History

1. This notion of the oceans as interactive spaces requires maritime historians to look beyond the **shoreline**. Maritime connections generated dynamics that affected not only coastal regions and **seafaring** communities, but also the **hinterland**, as the circulation of goods, people and ideas across the oceans **transformed** economies and brought political change. Michael Pearson, who opened the "Connected Oceans" conference, long ago challenged any rigid land/sea divide as a false **dichotomy**, stating that "a maritime historian has to face the question of how far inland must we go before we can say that the ocean no longer has any influence". It is in this spirit that this Forum places the **interconnectedness** of the sea and the land at the center of its investigation: in what ways did oceanic routes integrate coastal stretches as well as hinterlands into wider systems of interaction? What was the role of port cities and merchant communities in the formation of cross-oceanic networks of exchange? To what extent did environmental factors determine political or economic processes? And what were the implications of global economic changes on individual ports or regional port-systems?

2. The ways in which port cities were integrated into regional as well as global systems of exchange are also central to the essays of Filipa Ribeiro da Silva and Ana Catarina Garcia, which take us from the history of the Pacific and Indian Oceans to that of the Atlantic. Ribeiro da Silva examines major shifts in the configuration of the port systems of Atlantic Africa between the fifteenth and nineteenth centuries. Through an analysis of the volumes of shipping, she assesses the impact of the transatlantic slave trade, as well as other political and economic dynamics, on the various connected and interdependent ports along Africa's western **littoral**. Finally, Catarina Garcia provides a comparative study of four colonial ports in the Atlantic and Caribbean, giving

leeway to another trend in maritime history: the study of empire-building. Her article examines the various economic, military and topographic factors surrounding the founding and development of the Portuguese port towns Angra on the Azores and Funchal on Madeira, and the British ports of Bridgetown and Port Royal on Barbados and Jamaica, concluding that environmental factors were of crucial importance for the successful development of these imperial **outposts**.

3.　Another **facet** that unites the studies presented here is their attention to the interplay of local and global dynamics. It is in this sense that this Forum on "Connected Oceans" builds on Sanjay Subrahmanyam's notion of "connected histories", which considers the dynamic between "the local and regional" and "the supra-regional, at times even global" —a key characteristic of early modern history. Whether focusing on economic connections or cultural transfers, or a combination of the two, the five authors explore the **manifold** ways in which global processes manifested in different local **constellations**, or how they were constituted by them. In combining global and local perspectives, they thus **align** with Patrick Manning's view that "our research framework needs to account **explicitly** for the global as well as the regional level of experience".

4.　The topics explored in this Forum do not, of course, exhaust the recent trends and debates in the field of maritime history, as illustrated. This was also testified to by the diversity of research projects presented at the conference in Porto. Three of them are included in the form of research notes in this issue of the *International Journal of Maritime History*. Eberhard Crailsheim explores the social mechanisms through which merchant groups in early modern Seville and Manila established the level of interpersonal trust that allowed them to forge globe-spanning trade relations. Going in a very different direction, Ana María Rivera Medina presents "E-port". Atlantic Cartography, XIVth—XVIIIth centuries, a project in digital humanities that illustrates the benefits of using new technologies when working with historical maps as primary sources.

5.　Studying the changing cartographic representations of coastal areas and port cities in the Atlantic, the interdisciplinary "e-port" project has created a database of more than 800 ports, which is also available to other researchers under the open access scheme. José Luís Gasch-Tomás, Koldo Trápaga Monchet and Ana Rita Trindade—who all work on the research project "For Sea Discovery"—examine the shipbuilding policies of the Spanish Empire in the early seventeenth century. Their investigation is guided by the wider research questions of "For Sea Discovery": to what extent did shipbuilding cause **deforestation** in the Iberian Peninsula during the early modern era and how did it compare to other causes of forest clearance, such as agriculture or rising demands for **timber** due to population growth?

6. However, investigated here primarily as a zone of cultural encounters, as a means of economic interaction, as sublime space of human imagination, the sea also needs to be treated as an object of analysis in its own right. One particularly promising area for further research lies beneath the blue surface—in the connection between human activities and marine life, between maritime and environmental history. Historians and marine environmentalists alike complain about the almost complete lack of marine environmental history, particularly for the early modern period.

7. The social, ecological, and economic value of the ocean is beyond doubt. As a supplier of goods, it provides considerable amounts of food, minerals and energy resources. As a means of transportation, it facilitates trade, tourism and transfers. As a source of renewable energy, the oceans comprise a potential that mankind has just begun to exploit (tidal electricity generation, wave energy **converter** and ocean thermal energy conversion). Acknowledging the economic relevance of the sea justifies the increasing interest in studying it from an environmental point of view. Simultaneously, since humans have only recently acquired the technological capacity to systematically and exhaustively extract resources from the oceans, environmental concerns about the preservation of ocean resources also demand attention. These aspects have been dealt with by a number of research projects that deserve further mention.

8. One of the most important of those, and one of a few with a historical **trajectory**, is the "History of Marine Animal Populations", a global initiative to study past ocean life and how it was impacted by human interaction with the sea. The project forms part of the "**Census** of Marine Life", an interdisciplinary and international programme that has assessed and estimated the abundance, distribution and diversity of ocean life, past, present and future. Another example is the "**Estuarine** Living Marine Resources" project conducted by the US National Oceanic and Atmospheric Administration from 1985 to 2000 which developed a comprehensive database of 122 species—from fish and shellfish over migrating birds to coastal shore wildlife—inhabiting North-American **estuaries**. More recently, the "Oceans Past Platform" (OPP) brought to-

gether historians, archaeologists, social scientists and marine scientists to engage in dialogue and collaboration with ocean and coastal managers in order to develop a better understanding of the value of marine resource extraction to European societies.

9. From a historical point of view, the universe of topics in maritime or marine environmental history seems almost infinite: questions of biodiversity, marine ecosystems, consumption patterns, marine resource depletion, energy production and **hazardous** waste management are but some of the topics that might be explored. However, evaluating long-term changes and environmental processes is invariably a difficult, if not almost impossible, task. In order to **forge** new pathways of research, maritime historians will have to continue being creative in gathering new source material and extracting fresh information from existing evidence. And while historians will need to uphold their thorough contextual and textual analysis, they also will need to make available source metadata and create **calibrated** digital datasets.

10. This should go along with a willingness to use new methodologies made available by other disciplines, for example for means of modelling and visualization. Such a collaborative approach could allow for analyzing environmental and geological changes, along with the social and economic mechanisms responsible for these changes, and their long-term consequences. Testing hypotheses based on interdisciplinary methods and creating significant samples that can be processed by mathematical, ecological or geological modelling are some parallel strategies designed to overcome the difficulties environmental history faces due to the lack of consistent historical data. Establishing a functioning cooperation between the different disciplines is therefore paramount for such a set-up, and it presents one of the biggest challenges for maritime historians.

(1,383 words)

➤ *New Words*

shoreline [ˈʃɔːlaɪn]
    n. a boundary line between land and water 海岸线；海岸线地带
Over the years he and his colleagues have produced scores of research papers, atlases of shoreline change and other reports.
多年来，他和同事们已经撰写了大量研究论文、海岸线变化图册和其他报告。

seafaring [ˈsiːfeərɪŋ]
    a. used on the high seas 航海的
Diseases of the heart and arteries end many seafaring careers early, and are the number one killer of seafarers while at sea.
心脏和动脉疾病过早地结束了许多人的航海生涯，而且也是海员们的头号杀手。

hinterland [ˈhɪntəlænd]
    n. the area of land behind it or around it 内地
The Eurasian hinterland has always been the land passage linking the east with the west and was once one of the most vigorous regions in the world economic landscape.
亚欧大陆腹地历来是东西方之间的陆上通道，也曾是世界经济版图中最活跃、最发达的地区之一。

transform [trænsˈfɔːm]
    v. to change or alter in form, appearance, or nature 变换，改变
Your metabolic rate is the speed at which your body transforms food into energy.
你的新陈代谢率就是你身体将食物转换成能量的速度。

dichotomy [daɪˈkɒtəmɪ]
    n. being twofold 二分法；两分
A dichotomy may be defined as a Dedekind cut for a weak ordering.
二分法可以被定义为对弱排序的狄德金分割。

interconnect [ˌɪntəkəˈnekt]
    v. to be interwoven or interconnected 使互相连接；互相联系
All organizations face integration problems of some sort; perhaps because of a corporate merger, a new business alliance, or just the need to interconnect existing systems.
任何组织都会碰到某些类型的集成问题；或许是因为一次公司的合并、一个新的商业联盟或者仅仅是需要相互连接现有的系统。

littoral [ˈlɪtərəl]
    n. to the region of the shore of a lake or sea or ocean 沿海地区（等于 litora）
Along this narrow littoral, a century ago, the Tekke Turcomans had grazed their camels and tough Argamak horses, and tilled the soil around forty-three earthen fortresses.

一个世纪之前,在这个滨海地区的沿岸,提基亚土库曼人用泥土建立起 43 个堡垒,他们在周围放牧骆驼和凶悍的阿葛马克马,并耕种土地。

leeway['li:weɪ]　　　　*n.* a permissible difference; allowing some freedom to move within limits 自由活动的空间;余地

The UN resolution is wide-ranging, giving the coalition leeway not just to disable Col Gaddafi's air defences but also target Libyan ground forces.

联合国决议给联盟留下的回旋余地不仅是要重伤卡扎菲的防空系统同时也要摧毁利比亚的地面武装力量。

outpost['aʊtpəʊst]　　*n.* a station in a remote or sparsely populated location 前哨

The region, which was the target of a recent NASA mission that intentionally crashed an unmanned spacecraft to look for frozen water, could be ideal for a future human outpost.

这片区域最近成为 NASA 一项无人飞行任务的目的地,这次任务使用无人驾驶飞船撞击这一区域以检测是否存在固态的水。如果有的话,这里将成为人类设立前哨的理想地点。

facet['fæsɪt]　　　　*n.* a distinct feature or element in a problem 面;方面;小平面

Indeed, almost every facet of software development proffers at least one framework.

确实,几乎软件开发的每个方面都提供了至少一个框架。

manifold['mænɪfəʊld]　*a.* many and varied; having many features or forms 多种多样的,许多种类的

The moves in chess are not only manifold, but involute.

下象棋的走法不但多种多样,而且错综复杂。

constellation[ˌkɒnstə'leɪʃn]　*n.* an arrangement of parts or elements 星座;星群

But with the cancellation of NASA's Constellation programme to return Americans to the moon by 2020, who is to inspire the next generation?

但是随着美国国家航空航天局 2020 年重返月球的"星座计划"取消,未来还会由谁来推动下一代人完成这一计划呢?

align[ə'laɪn]　　　　*v.* to change something slightly so that it is in the correct relationship to something else 使一致

Domestic prices have been aligned with those in world markets.

国内价格已调整到与世界市场一致。

explicitly[ɪk'splɪsɪtlɪ]　*ad.* in an explicit manner 明确地;明白地

These requirements and extortions explicitly violate prohibitions of the World Trade Organization.

那些要求和勒索明确地违反了世界贸易组织的禁令。

deforestation [ˌdiːˌfɔːrɪˈsteɪʃn] *n.* the state of being clear of trees; the removal of trees 毁林,采伐森林,烧林

The report's findings, raising the profile of peat, contrast with the conclusions of a paper on deforestation published this week in *Nature Geoscience*.

报告结果提升了泥炭的关注度,这与本周在《自然地球科学》上发表的一片关于森林砍伐的文章形成了对比。

timber [ˈtɪmbə(r)] *n.* the wood of trees cut and prepared for use as building material 木材;木料

He logged the timber into 3 foot lengths.

他将木材锯成 3 英尺长的段木。

converter [kənˈvɜːtə(r)] *n.* a device for changing one substance or form or state into another 转换器;整流器;变频器

We will create a custom message converter to send a trade creation request message from the client to the middle tier.

我们将创建一个自定义消息转换器,来发送一个从客户端到中间层的关于交易创建的请求消息。

trajectory [trəˈdʒektərɪ] *n.* the path followed by an object moving through space 轨道,轨线

When we throw up an object in a gravitational field, we'll get the trajectory in a plane.

当我们把一个物体放到引力场中,我们会得到在一个平面上的轨迹。

census [ˈsensəs] *n.* a periodic count of the population or else (官方的) 统计,调查

That is the big message which will be coming from the census on top of its maps of ocean biodiversity and the new understanding of what lives where and why.

基于海洋生物多样性地图和人们对海洋生物的分布及其机理的最新认识,人们将从这个海洋调查中获得上述的重要讯息。

estuarine [ˈestjuərain] *a.* of or relating to or found in estuaries 河口的,江口的

Zookeepers said they now have a population of 80 red-eared turtles and that they were forced to build a new enclosure for the dozen estuarine crocodiles found last month.

动物园管理员说,园内现有 80 只红耳龟,他们不得不修建新的围栏用来圈养数十只上个月被发现的河口鳄鱼。

estuary [ˈestʃuərɪ] *n.* the wide part of a river where it nears the sea; fresh and salt water mix 河口;江口

Fresh water from the Parana and Uruguay rivers collides with salt water from the South Atlantic Ocean in the muddy estuary of Rio de la Plata.

巴拉那河和乌拉圭河的淡水与来自南大西洋的海水在拉普拉塔泥泞的河口相互碰撞着。

hazardous [ˈhæzədəs]   *a.* involving risk or danger 有危险的;冒险的

Coupled with better communication, as the human population skyrockets and we move into more hazardous regions, we're going to hear more about the events that do occur, Arrowsmith added.

阿罗史密斯补充道,由于人口的迅速增加,我们迁移到更危险的地区,再加上通信越来越发达,我们将听到更多类似事件发生的消息。

forge [fɔːdʒ]   *v.* to come up with (an idea, plan, explanation, theory, or principle) after a mental effort 缔造

The prime minister is determined to forge a good relationship with the country's new leader.

该首相决意与该国的新领袖建立良好的关系。

calibrate [ˈkælɪbreɪt]   *v.* to make fine adjustments or divide into marked intervals for optimal measuring 校正

From the Moon, Earth is an ideal reference point to calibrate the cameras.

从月球的角度看,地球是一个理想的校准相机的参照点。

➤ *Phrases and Expressions*

a combination of 事物的结合;综合……
as illustrated 如图所示
beyond doubt 毋庸置疑,确实地
point of view 观点;见地;立场
go along with 赞同;陪……一起去

➤ *Proper Names*

Madeira 马德拉群岛(大西洋的群岛名)
Barbados 巴巴多斯(拉丁美洲国家)
Jamaica 牙买加(拉丁美洲一个国家)
Porto 波尔图(葡萄牙港口)
the Iberian Peninsula 伊比利亚半岛
Oceans Past Platform (OPP) 海洋通过平台

➤ *Translation*

1. Maritime connections generated dynamics that affected not only coastal regions... (Para. 1)

   随着跨越海洋的货物、人员和思想的流通改变了经济并带来了政治变革,海洋联系产生的动力不仅影响到沿海地区和航海区域,而且也影响到内陆地区。

   【这个句子较长,原封不动地对原句进行从前到后地翻译,译文会让读者感到费解。将原文的后半句前移,调整语序,采用把句子断开来翻译的断句译法进行翻译,译文更能为读者接受。】

2. Through an analysis of the volumes of shipping... interdependent ports along Africa's western littoral. (Para. 2)

   通过对货运量的分析,她评估了跨大西洋奴隶贸易以及其他政治和经济动态对非洲西部沿海地区相互联系和相互依赖的各个港口的影响。

   (这里主要是指《国际海事史杂志》公布的调查研究。)

3. Another facet that unites the studies presented here is their attention to the interplay of local and global dynamics. (Para. 3)

   另一方面,他们对地方和全球动态的相互作用的关注也将这里的研究联系在一起。

4. The topics explored in this Forum do not, of course... in the field of maritime history, as illustrated. (Para. 4)

   当然,本论坛所探讨的主题并没有像所表明的那样,详细说明海洋历史领域的最近趋势和争论。

5. Studying the changing cartographic representations... available to other researchers under the open access scheme. (Para. 5)

   这个跨学科的"电子口岸"项目研究了大西洋沿岸地区和港口城市不断变化的地图表现形式,创建了一个包含 800 多个港口的数据库,其他研究人员也可以在开放存取计划下使用这个数据库。

6. However, investigated here primarily as a zone of cultural encounters, as a means of economic... ic... (Para. 6)

   然而,在这里主要作为一个文化接触区、一种经济相互作用的手段、人类想象的崇高空间来研究,海洋本身也需要作为一个分析对象来看待。

7. The social, ecological, and economic value of the ocean... minerals and energy resources. (Para. 7)

   海洋的社会、生态和经济价值是毋庸置疑的。作为货物的供应者,它提供了大量的食物、矿物和能源。

8. These aspects have been dealt with by a number of research projects that deserve further mention. (Para. 7)

   一些值得进一步提及的研究项目已经处理了这些内容。

9. And while historians will need to uphold their thorough... and create calibrated digital data-sets. (Para. 9)

虽然历史学家需要坚持进行彻底的上下文和文本分析,他们也需要提供可用的源数据和创建校准的数字数据集。

10. Establishing a functioning cooperation... and it presents one of the biggest challenges for maritime historians. (Para. 10)

因此,在不同学科之间建立起有效的合作关系,对于此类机构至关重要,这也是海事历史学家面临的最大挑战之一。

【这个句子句式结构不太复杂,但在翻译时注意英汉语法结构上的差异,采用换序译法改变原文词语的前后次序,使译文通顺自然,符合汉语的表达习惯。将"therefore"译为"因此"置于句首,将介词短语"between the different disciplines"译为"在不同学科之间"放在动词"建立"的前面,将"for maritime historians"译为"海事历史学家"置于"面临"的前面,做到译文的自然通顺。】

Exercises

1. Fill in the blanks with the proper given words, and then translate the sentences into Chinese.

transform	comprise	infinite	visualization
geological	seafaring	assess	topographic
testify	marine	interdisciplinary	consumption
manifest	outpost	forge	sublime

1) They can work with the project teams to _____ and understand needs, and then expedite the search, understanding, and analysis of services and guide teams in how and when to use them.

2) Since I don't want to extend too much with the overall architecture description, I prefer to stop on my preferred topics and subsystems that _____ the project.

3) China is a country with unique _____ and geomorphic conditions and complicated engineering geology, resulting in extremely complex technical problems in building transportation infrastructure.

4) A moderate increase in both investment and _____ will play an active role in stimulating the market.

5) Soldiers manning a remote _____ in northern Helmand have been forced to go foraging for food between attacks by the enemy.

6) This not only provides you with another form of _____, it also uses a formal, technical

model underneath.

7) But even this method runs the risk of altering the acidity of soil or _____ ecosystems unless precautions are taken with disposal.

8) In the real world, we validate signatures by sight, even though a skilled attacker can reliably _____ a signature that most people cannot distinguish from the original.

9) When John Lennon put pen to paper, the result was usually inspired and occasionally _____. But not always.

10) Today, we need to be able to perform _____ searches and, crucially, be able to acquire the material from a vast number of sources.

11) Europe faces all sorts of social and political problems, which may or may not _____ themselves, but its current dilemma is rooted in basic economics.

12) Auroras, such as the one pictured above, pleasingly _____ to a stream of particles from the sun that gets through and hits the atmosphere.

13) CBS' "60 Minutes" produced a remarkable story explaining how the _____ Moken people used their close connection to the ocean to detect the tsunami before it happened.

14) Geologists care about sediments, hammering away at them to uncover what they have to say about the past-especially the huge spans of time as the Earth passes from one _____ period to another.

15) The function accepts three arguments, and performs some simple tests on the values to ensure that the function will not start an _____ loop.

16) Cell stems are so powerful because they have the amazing ability to _____ and grow into other types of cells—such as blood cells or bone cells.

2. Translate the following passage into Chinese.

One of the most important of those, and one of a few with a historical trajectory, is the "History of Marine Animal Populations", a global initiative to study past ocean life and how it was impacted by human interaction with the sea. The project forms part of the "Census of Marine Life", an interdisciplinary and international programme that has assessed and estimated the abundance, distribution and diversity of ocean life, past, present and future. Another example is the "Estuarine Living Marine Resources" project conducted by the US National Oceanic and Atmospheric Administration from 1985 to 2000 which developed a comprehensive database of 122 species—from fish and shellfish over migrating birds to coastal shore wildlife—inhabiting North-American estuaries. More recently, the "Oceans Past Platform" (OPP) brought together historians, archaeologists, social scientists and marine scientists to engage in dialogue and collaboration with ocean and coastal managers in order to develop a better understanding of the value of marine resource extraction to European societies.

3. Translate the following passage into English.

港口城市融入区域和全球交换体系的方式,也是菲利帕·里贝罗·达席尔瓦和安娜·卡塔琳娜·加西亚文章的中心内容。这两篇文章将我们从太平洋和印度洋的历史带到大西洋的历史。里贝罗·达席尔瓦考察了 15 到 19 世纪期间大西洋非洲港口系统结构的重大变化。通过对货运量的分析,她评估了跨大西洋奴隶贸易以及其他政治和经济动态对非洲西部沿海地区相互联系和相互依赖的各个港口的影响。最后,卡塔琳娜·加西亚对大西洋和加勒比海的四个殖民港口进行了比较研究,为海运史上的另一种趋势提供了余地:建立对帝国的研究。

Unit 2
Marine Literature

Aquaman

1. Stalnoivolk, a six-hundred-foot Russian submarine, churned through the waters a few miles off the eastern Atlantic **coastline**, the seven blades of its propeller cutting through the sea like an underwater ferris wheel. Attached to the side like a suckerfish was a **sleek**, high-tech miniature submarine, its white running lights flashing across its pitch-black **hull**. As sea life parted to get out of the massive sub's path, inside, the crew was seeking shelter from rapid machine-gun fire.

2. The emergency lights inside the control room cut out as the machine gun ceased fire and the alarms went silent. A middle-aged African-American man in a **slick** wet suit hoisted the gun to his shoulder, the barrel still smoking, as he stepped over the bodies of the dead Russian crewmen. Jesse Kane gave a wicked grin as he watched his band of pirates take their places at the ship's controls.

3. "Distress signal and all alarms have been disabled," came a voice from behind Jesse. He turned to see a muscular, broad-shouldered younger version of himself, dressed in a heavily modified wet suit. The younger man's smile was even more dangerous than Jesse's, if possible. Jesse knew it was possible, because the man was his own son. "Excellent, David. We're running dark again." Jesse nodded to the crewman behind David, signaling them to bring their captive forward. "But they heard it, of that you can be sure, American." The sub's captain was bloodied and beaten, yet still his voice was raised in **defiance**. This was a man who knew he was facing certain death, but who would fight to the bitter end, ready to go down with his ship. Jesse admired that in the man.

4. David sneered at the captain. "Make you a deal. I won't tell you how to captain, you

· 306 ·

don't tell me how to pirate." He flicked his wrist and a long, **lethal** blade extended from the forearm of his suit. "On second thought, I'm not in the mood for deals today, so consider yourself relived of duty." As David wiped the blood from his blade, the captain's lifeless body fell the floor, a **gash** cleanly cut through the front and back of his **torso**. Jesse was proud of the way his son took charge. He motioned for David to follow him as he walked to the officers' quarters at the bow of the ship.

5. David looked around at the empty **bunks**. "Where's the rest of the crew?" "It seems our reputation precedes us. They've **barricaded** themselves near the torpedo bay ahead." Jesse reached into a pouch. "Or, I should say, your reputation, this porch." "Or, I should say, your reputation. This was your op, your win, and I think you've earned this." David looked at the item in his father's hand: a well-worn hunting knife. "I can't take that piece of junk. That's the love of your life. I've never seen you a day without it—sharpening that blade."

6. "It hasn't always been mine," Jesse said, placing the knife firmly into his son's hand. "It belonged to your grandfather. He was one of the navy's first frogmen in World War II and the first black man to have that honor. His fellow mates nicknamed him the **manta** for how stealthily he moved through the water... and how quickly he could kill a man armed just with his knife." David turned the blade over and saw an image of a manta ray **embossed** into the handle. "He gave it to me when I was your age. I think he'd want you to have it now. Carry on the tradition."

7. David was about to thank his father when something thudded hard against the top of the sub, almost making them lose their footing. Jesse activated the bluetooth communicator in his earpiece, "What hit us?". In the control room, one of the pirates was looking at the radar screen, his eyes not believing what they were seeing, "Sir... there's someone out there!" "Another sub?" Jesse was confused. Even if help was coming, no vessel could have reached them this quickly. "I think... it's a man!". Before anyone could react, the **behemoth** submarine began to rise toward the surface! "I gave no order to change course!" Jesse barked.

8. With an amazing splash, the submarine broke the surface of the water and rose another ten feet into the air before crashing back down, floating like a lame, oversized metal duck. Inside, the pirates held their breath as they heard footsteps above them, someone was walking on the hull! A **gleam** in his eye, David turned to his father, "That's not a man." "You think it's the 'Nessie' you've been chasing?" Jesse

asked. Before David could answer, there came the screening sound of metal being ripped open. Jesse's men rushed to the sound. The top hatch was missing! They raised their guns tentatively, whispering among themselves. Without warning, the hatch flew down into the sub, knocking two pirated to the ground, a figure dropped from the hull into the sub.

9. Indeed, there was no man—no ordinary man at least, towering above them with shoulder-length curly brown hair, golden eyes, and a bare broad chest covered in intricate **tattoos**, the mystery man stretched to his full height and gave a **smirk**. "Permission to come aboard", said Arthur Curry. The rat-a-tat-tat of gunfire form the pirates gave their answer. Arthur grabbed the hatch from the floor to use as a shield as he barreled down the subs hallway. Bullets **ricocheted** as he continued his march like a football player headed for the end zone. He stopped in front of a cabin and tore the door off as though it were paper. Turing to the last remaining pirate, he grinned. "Hold the door, will ya?" He asked as he threw it at the pirate, flattening him. Arthur stepped into the room. It was the **torpedo** bay where the surviving Russian crew had hidden.

10. "He—he's real!" said one in Russian. "And he's late for happy hour, so hurry up", Arthur answered, also in Russian. He led them up the stairs to the top of the submarine, where they quickly boarded three lifeboats. "Stay here, just need to take out the trash." With that, Arthur dropped back into the sub and found himself face-to-face with Jesse and David, the elder pointing his machine gun and the son with a glock trained on the intruder's chest. The duo opened fire. As they emptied their weapons, round after round knocking Arthur farther back, they drove him deeper toward the torpedo bay, to escape the shower of bullets, Arthur dropped through a hatch in the floor, deeper into the bowels of the sub.

11. Arthur had only a moment to notice he was surrounded by the ship's torpedoes that lined the walls. He caught a glimpse of an opening that had been cut with laser-sharp precision, beyond looked to be a control panel. The mini submarine was docked there, attached securely so as not to let in any water. The thunk of another man dropping into the torpedo bay shifted Arthur's attention, it was the younger of his attackers. The man triggered a long blade that sprang from his suit's forearm. He gripped a short sword in his other hand, **twirling** it slightly. This attacker had a look of glee on his face as he started toward Arthur. "Been waiting a long time for this,"

David said as he lunged. Arthur **dodged** the blades deftly, "and yet this is the first time I've seen you, couldn't have been waiting too long."

12. David engaged Arthur, crashing both blades down on him. Arthur blocked the move with a pipe he tore form the wall. The men's faces were inches apart. Arthur could see an odd excitement in the stranger's eyes. "I **scavenge** the seas and you patrol them, right, 'Aquaman'? We were bound to meet sooner or later," David welcomed the challenge to face the creature he had studied for years, scouring reports and news about sightings of the mystery man from the sea. This indeed was his "Nessie", and he meant to make Arthur a prized **trophy** when he killed him. "Huh, well, let's not make a habit of it, then", Arthur said, pushing David back and dropping the pipe.

13. David slashed again and again at Arthur, who blocked the blades with his bare arms, knocking them aside without so much as a scratch. David's eyes widened as Arthur swung his forearm and knocked the short sword from David's grip. Dodging a swipe from the man's wrist blade, Arthur caught it between his hand and snapped it in two. David found himself hoisted in the air and Arthur flung him across the way. David's body slammed into the steel wall and slid down.

(1,471 words)

➤ *New Words*

Aquaman [ˈækwɔːmən]   *n.*   a Filmation animated series that premiered on CBS《海王》(电影名);海王,水行侠(漫画人物)

Now technology can help mere mortals breathe underwater like Aquaman, repel bullets like Superman, and even manipulate the weather like Storm.

现在,技术能够帮助更多的凡人像海王一样在水下呼吸、像超人一样阻挡子弹,甚至像风暴女一样操控天气变化。

coastline [ˈkəʊstlaɪn]   *n.*   the outline of a coast 海岸线

Since there are so many rivers and so much coastline in the region, there is much scope for expanding fish-farming and seafood production.

因为该地区拥有如此多的河流和如此多的海岸线,扩大渔业和海产品生产拥有很大空间。

sleek [sliːk]   *a.*   having a smooth, gleaming surface 圆滑的

It was the biggest crow she had ever seen, plump and sleek, with rainbows shining in its black feathers.

这是她见过的最大的一只乌鸦,它丰满而且光滑,黑色的羽毛闪烁着彩虹般的光芒。

hull [hʌl]   *n.*   the frame or body of ship 船体

Almost all cargo vessels, by contrast, are flat bottoms, which allow a larger volume to be kept buoyant for a given amount of hull metal.

相比之下,几乎所有商船的底部都是平的,因为这样可以为庞大的金属船壳提供更大的浮力。

slick [slɪk]   *a.*   having a smooth, gleaming surface 光滑的

Boat wakes appear as slightly curved lines, accentuated by the pastel tones of the oil slick.

船尾呈略微弯曲的线条,浮油的柔和色调突显了这种感觉。

defiance [dɪˈfaɪəns]   *n.*   intentionally contemptuous behavior or attitude 反抗

Angry protestors with clenched fists shouted their defiance.

愤怒的抗议者们紧握拳头高声抗议。

lethal [ˈliːθl]   *a.*   of an instrument of certain death 致命的,致死的

Scientists haven't yet figured out what causes the mutations, when they will occur and what makes certain viruses more lethal than others.

科学家迄今尚未找出突变的原因,它们什么时候会发生及为什么

某些病毒比其他病毒更致命。

gash[gæʃ]   *n.* a strong sweeping cut made with a sharp instrument 很深的裂缝；砍得很深的伤口

There was an inch-long gash just above his right eye.

就在他的右眼上方，有一道一英寸长的伤口。

torso['tɔːsəʊ]   *n.* the body excluding the head and neck and limbs 躯干

The swimmer who makes the bigger wave is the faster swimmer, and a longer torso makes a bigger wave.

游泳者制造的浪越大，就游得越快，躯干越长，则制造的浪就越大。

bunk[bʌŋk]   *n.* a rough bed (as at a campsite) 铺位；床铺

Getting to that top bunk demanded agility; it was reached by a very vertical, very narrow ladder.

爬到上面铺位需要身手敏捷，那里有一架狭窄的垂直梯子。

barricade['bærɪkeɪd]   *v.* to defend or block something by building a barricade 设路障防护；阻挡

They barricaded all the doors and windows.

他们用障碍物堵住了所有的门窗。

manta['mæntə]   *n.* any large ray (fish) of the family Mobulidae, having very wide wing like pectoral fins and feeding on plankton 魔鬼鱼

We see the same pattern—now with these tags we're seeing a similar pattern for swordfishes, manta rays, tunas, a real three-dimensional play.

通过这些标记，我们在旗鱼、魔鬼鱼、金枪鱼等身上也看到了同样的行为模式，一个真正三维立体的活动。

emboss[ɪm'bɒs]   *v.* to mould or carve (a decoration or design) on (a surface) so that it is raised above the surface in low relief 在……上做浮雕图案；装饰

Now, we are going to add a glass effect to the copy text layer (the text layer on the top), so double click it, and add a bevel and emboss effect.

现在，我们要复制文本层（在顶部的文本层）添加玻璃效果，然后双击此处，添加斜面和浮雕效果。

behemoth[bɪ'hiːməθ]   *a.* huge, tremendous 巨大的；高大的

Compared to the behemoth task of manually upgrading a user base with a locally installed application, the effort required to upgrade users with a thin client application is minimal.

相对于本地安装的应用程序需手动为用户群进行升级的巨大工程而言，利用瘦客户应用程序为用户升级的工作量要少得多。

gleam[gliːm]   *n.* an appearance of reflected light 微光；闪光

Their hope did not last long, and the gleam was quickly eclipsed.

希望没有延长多久，微光很快就消逝了。

Nessie['nesɪ]　　　　*n.* a large aquatic animal supposed to resemble a serpent or plesiosaur of Loch Ness in Scotland 尼斯湖水怪

Years passed, and evidence proving Nessie's existence (including photographs, video and even sonar) mounted.

很多年过去了，有证据表明尼斯湖水怪真的存在，证据包括照片、视频，甚至是声呐系统。

tattoo[tə'tuː]　　　　*n.* a design on the skin made by tattooing 文身

He had a tattoo on his arm representing his family.

在他胳膊上有一个代表着他家族的文身。

smirk[smɜːk]　　　　*n.* a smile expressing smugness or scorn instead of pleasure 傻笑，得意的笑；假笑

His eyes flickered up at me from under his lashes, the hint of a smirk on his face.

他的眼睛从睫毛下飞快地看了我一眼，脸上露出一丝坏笑。

ricochet['rɪkəʃeɪ]　　　　*v.* to hit a surface and come off it fast a different angle 弹开；反弹出去

Increasingly, an event in one part of the world can quickly ricochet throughout the international system to affect us all.

世界任何一地的事件越来越迅速地在整个国际系统引起反响，影响到我们每个人。

torpedo[tɔː'piːdəʊ]　　　　*n.* armament consisting of a long cylindrical self-propelled underwater projectile that detonates on contact with a target 鱼雷，水雷

During the 3-day Battle of Tarawa, some 1,000 US Marines died, and another 687 US Navy sailors lost their lives when the USS *Liscome Bay* was sunk by a Japanese torpedo.

持续3天的塔瓦拉战役，约1 000名美国海军陆战队员丧生；由于美军护航航母"利斯康姆湾"号被日军鱼雷击沉，另有687名水兵丧生。

twirl[twɜːl]　　　　*v.* to turn in a twisting or spinning motion 使轻快地转动

When I twirl this around, it produces a particular tone.

当我把这个旋转时，它会发出特殊的音调。

dodge[dɒdʒ]　　　　*v.* to make a sudden movement in a new direction so as to avoid 躲开；迅速让开

He ran across the road, dodging the traffic.

他躲开来往的车辆跑过马路。

scavenge['skævɪndʒ]　　　　*v.* to remove unwanted substances from 打扫；清除污物

In the nearest future the equipment will scavenge the sound produced by motorway traffic at rush hour and using it to give the national grid a boost.

在不久的将来,这种装置也可用来清除汽车高速公路高峰时产生的噪声,并将产生的电输送给电网。

trophy[ˈtrəʊfɪ]   *n.* an award for success in war or hunting 奖品;战利品;纪念品

With two sides from the same nation competing for the trophy, it was always likely to be a cagey affair.

当两支来自同一个国家的球队竞争奖杯的时候,往往会是一场小心翼翼的决斗。

➤ *Phrases and Expressions*

attached to 附属于;系于
if possible 如果可能的话;如有可能
on second thought 进一步考虑后,仔细考虑后
at the bow of 在船头
a glimpse of 瞥见
sooner or later 迟早,早晚

➤ *Proper Names*

Stalnoivolk 钢铁狼(俄罗斯潜艇)
ferris wheel 摩天轮

➤ *Translation*

1. Attached to the side like a suckerfish was a sleek, high-tech miniature submarine, its white running lights flashing across its pitch-black hull. (Para. 1)

像胭脂鱼一样贴在船侧的是一艘光滑的高科技微型潜艇,白色的航行灯在漆黑的船体上闪烁。

【由于英语和汉语语法结构上的差别,翻译中把这个句子的"attached to..."放在"like a suckerfish..."后面翻译,采用改变原文词语的前后次序的换序译法翻译更为妥当,译文通顺易于理解。】

2. A middle-aged African-American man in a slick... over the bodies of the dead Russian crewmen. (Para. 2)

一名穿着光滑潜水衣的非洲裔中年美国人把枪举到肩上,枪管还冒着烟,跨过俄罗斯船员的尸体。

3. As David wiped the blood from his blade... front and back of his torso. (Para. 4)

当大卫擦拭着刀刃上的血迹时,船长毫无生气的身体倒在了地板上,一条清晰的伤口划破了他的前胸和后背。

4. It seems our reputation precedes us. They've barricaded themselves near the torpedo bay ahead. (Para. 5)

看来我们是名过其实了。他们在前面鱼雷舱附近设置了路障。

5. His fellow mates nicknamed him the manta for how stealthily... he could kill a man armed just with his knife. (Para. 6)

他的同伴们给他起了个绰号叫"魔鬼鱼",因为他在水里游得非常快,而且他能很快杀死一个拿着刀的人。

6. David was about to thank his father when... against the top of the sub, almost making them lose their footing. (Para. 7)

大卫正要感谢他的父亲,突然有个东西重重地撞到了潜艇的顶部,几乎使他们失去了平衡。

7. With an amazing splash, the submarine broke... floating like a lame, oversized metal duck. (Para. 8)

随着一声惊人的溅水声,潜艇浮出水面,又升了10英尺,然后又坠落,像一只跛脚的、特大号的金属鸭子一样漂浮着。

8. Indeed, there was no man—no ordinary man... the mystery man stretched to his full height and gave a smirk. (Para. 9)

确实,没有一个人至少不是一个普通人,高高在上,长着齐肩的棕色卷发,金色的眼睛,裸露的宽阔的胸膛上布满了复杂的文身,这个神秘人伸直了身体,露出得意的笑容。

9. As they emptied their weapons... through a hatch in the floor, deeper into the bowels of the sub. (Para. 10)

随着子弹的消耗,他们一轮又一轮地把亚瑟打得越来越远,他们把他往鱼雷舱的深处里推。为了躲避枪林弹雨,亚瑟从地板上的一个舱口掉了下去,掉到了潜艇内部的深层处。

【这个句子比较长,都是由逗号分成的短句。将后半句的动词不定式结构与后面短句转成汉语的一个句子的转句译法进行翻译,译为"为了……"。这个含义丰富的短句就被简洁明了地表达出来。】

10. David slashed again and again at Arthur... without so much as a scratch. (Para. 13)

大卫一遍又一遍地猛砍亚瑟,亚瑟光着胳膊挡住了刀刃,把刀刃弹到一边,胳膊上却没有留下一点伤痕。

Exercises

1. Fill in the blanks with the proper given words, and then translate the sentences into Chinese.

slash	stretch	trigger	sleek
pirate	blade	submarine	defiance
torpedo	intruder	glee	grip
scratch	flatten	relive	bare

1) A second way is also quite intuitive. You may be willing to pay more for a cashmere sweater or a small, _____ smart phone just because you like how it feels.

2) The problem for NATO is whether such hints of the old, cold-war posture might not _____ confrontation with Russia.

3) Ushakov envisaged his craft flying ahead of the target, landing on the sea and then flooding its fuselage so that it could lie in wait beneath the surface and _____ the ships as they sailed past.

4) In every community, even the crew of a _____ ship, there are acts that are enjoined and acts that are forbidden, acts that are applauded and acts that are reprobated.

5) And while this pain is normal, sometimes that pain lingers for too long. We _____ the pain over and over, and have a hard time letting go.

6) After that, that was constantly thrown back at me as Iraq deteriorated: "you were an optimist and now the security forces you developed are in the _____ of the militias", or this or that.

7) They also suggest that the sounds may function to attract the attention of predators, in which case the _____ would be rather defenseless.

8) For budgetary reasons, the redevelopment of these systems from _____ can be a poor option.

9) His call is bracing but melodious, although once he sets to work on a _____, the noise of grindstone on metal brings out the old women with their beloved worn-out cleavers.

10) With only the most rudimentary of equipment, anyone in the world can relay anything in the world, a death toll, an explosion, a cry for help, an utterance of _____.

11) Whether you are anticipating tomorrow's technology with _____, worry, or a combination of the two, you will be better informed and better able to take part in the connected world if you read this book.

12) "On the stairs," Tyler and Marla _____ against the wall as police and paramedics charge

by with oxygen, asking which door will be 8G.

13) It was about the period that my narrative has reached, a bright frosty afternoon; the ground _____, and the road hard and dry.

14) An optimist might wonder if Europe is about to embrace structural reform by accident: after allowing public sectors to swell, the need for swift cuts is pushing them to _____ the costs of the state.

15) For years, the Soviet navy released nuclear waste into the sea, including several spent _____ reactors that were dropped overboard at undisclosed locations.

16) It is not too much of a _____ to relate the deflated build-up to the World Cup in England this time to the wider, sombre atmosphere.

2. Translate the following passage into Chinese.

Arthur had only a moment to notice he was surrounded by the ship's torpedoes that lined the walls. He caught a glimpse of an opening that had been cut with laser-sharp precision, beyond looked to be a control panel. The mini submarine was docked there, attached securely so as not to let in any water. The thunk of another man dropping into the torpedo bay shifted Arthur's attention, it was the younger of his attackers. The man triggered a long blade that sprang from his suit's forearm. He gripped a short sword in his other hand, twirling it slightly. This attacker had a look of glee on his face as he started toward Arthur. "Been waiting a long time for this," David said as he lunged. Arthur dodged the blades deftly, "and yet this is the first time I've seen you, couldn't have been waiting too long."

3. Translate the following passage into English.

几百英里以外的海底深处,在距离两个来自不同世界的人第一次相遇的地方另一次相遇正在发生。令人叹为观止的建筑一片片地躺在海底。尽管这些遗迹已经支离破碎,苍白无力,但比起自"亚特兰蒂斯"号沉没以来的几个世纪里人类所建造的任何东西,这些遗迹都更加引人注目。正是在这些废墟中,来自不同王国的两支分遣队决定走到一起,举行一次意义重大的峰会。至少,这是一位领导人所希望的。这名男子骑着一头巨大的暴龙(一种大多数人认为已经灭绝的史前动物),挺直了腰板。

Unit 3
Marine Geography

Geographies of Oceanic Overproduction

1. **Scrap** data since 1975 shows that both progressive and periodic phases of devaluation have not only produced increasing quantities of devalued fixed capital but have become part of the logic of capital circulation through the shipping sector. This new scale of devaluation reveals how overproduction is, in part, produced by overaccumulated money capital moving into the shipping sector. This orientation of capital could not "be accomplished without a money supply and credit system which creates 'fictional capital' in advance of actual production and consumption". It is evident that each shipping crisis was preceded by an influx of increasing amounts of **fictitious** capital used to finance new technologies. These innovations sped up capital's **turnover** time in and through the shipping sector. This led to capital being advanced faster than it could be absorbed. This brings us to the increasing use of shipbreaking and laying up ships as strategies for spatially displacing economic losses and helping bring about new cycles of accumulation—with corresponding implications for labor and the environment.

2. Bangladesh and shipbreaking. On shipbreaking beaches in Chittagong, laborers, barefoot or in **flip-flops**, navigate a coastal landscape of roughly cut steel plates **haphazardly stacked**, fragments of broken pieces of machinery, uneven piles of wires, metal pipes and tubes, and forms of material waste such as **asbestos**, polychlorinated biphenyls (PCBs), and polycyclic-aromatic hydrocarbons (PAHs) compound. Gas and oil **residues linger** in enclosed areas on beached ships and often cause explosions when cuttermen light their torches. Accidents from gas explosions or workers being fatally struck by falling pieces of ships are widely reported in local and international media. The exact number of people injured or killed while working as shipbreakers is difficult to assess, because many accidents and fatalities go unreported in efforts to conceal the human stakes of the business. The NGO Shipbreaking Platform has documented at least 181 deaths between 2005 and 2017 in Bangladesh alone. In addition to

"spectacular" forms of violence, "slow violence" inflicted on workers manifests over time. Numerous reports have attributed long-term exposure to toxic materials to terminal and **debilitating** illnesses. The last reported fatality at the yard depicted was in September 2016, when an iron plate fell, struck, and killed a man working below. Jobs in shipbreaking are some of the most dangerous in the world. In 2016, Bangladesh's shipbreakers broke 8,240,000 million GT, absorbing 27.5% of the world's devalued **tonnage**.

3. The shipping sector's economic **woes** of the 1970s continued into the 1980s. In March 1981, shipping markets started a second deep slump, and overcapacity of tonnage grew. By 1985, the global scrap markets absorbed 22.3 million GT. This situation was brought about by a confluence of factors: the increase in variable interest rates on shipping **mortgages**, rising to 17%–20%; the impact of the OPEC oil price increase of 1979; and the fact that shipowners **exacerbated** an already weak market by ordering new ships to keep books high during the crisis. In 1981, orders for bulk carriers had swelled to 29 million DWT, or almost half of the total on order. Investor and financing strategies developed to include "asset play", fueled by surpluses in Asian dollar markets and the development of "**equity**-financing funds" from 1984 onwards. This was pioneered by Stokes' firm Bulk Transport Ltd. By the mid-1980s, the shipping sector suffered from continued overproduction as investors took advantage of market conditions, keeping second-hand ships active and ordering new builds. Some of the most prominent shipowners fell victim to their own speculative behavior. Between 1982 and 1987, over 100 million GT was scrapped, surpassing any previous five-year period. As devalued tonnage continued to plague the shipping sector, access to scrap markets became critical. Bangladeshi end-buyers bought ships from owners in Singapore with financing arranged through national banks in Bangladesh, and what began with five ships in 1981 grew to 59 in 1984. By 1988, Bangladesh emerged as one of the top three shipbreaking markets, buying 500,000 GT, almost 10% of the ships scrapped worldwide.

4. Consequences for labor and the environment were huge. In 1989, shipbreaking beaches covered 3.45 km. This doubled to 6.95 km in 1999 and 12.78 km in 2010. In order to expand shipbreaking capacity, end-buyers encroached upon coastal fishing villages and small-scale agricultural production, and they clear-cut coastal **mangrove** forests. Hossain highlighted the hazardous environmental and labor conditions caused by toxic materials released during shipbreaking through dumping, spilling, mixing, storing, burning, torch cutting, and moving. Laborers and the environment are exposed to liquid wastes (salt, acid, paint, **ammonia**, and **caustic** agents), heavy metals (lead, mercury, and **chromium**), PAHs, and PCBs as well as high amounts of asbestos. Nevertheless, for the government, the domestic "development" context will con-

tinue to outweigh the environmental and labor concerns as long as the steel extracted from this practice remains cheaper than steel imports. In maintaining one of the world's most important shipbreaking markets since the early 1990s, helping "mop up" overproduced and devalued fixed capital, Bangladeshi shipbreaking has been a vital market for the continued circulation of capital through the transport sector and the production of surplus value in the shipping economy. But with the ever-increasing build-up of surplus ships, competition-driven shipowners have also needed other places and practices to help mop up devalued vessels.

5.   Singapore's OPL (outer port limits). Singapore is a one-stop service center for maritime activities: brokerage and trading firms, ship financing and insurance services, **bunker**, crew and water suppliers as well as ship repair and marine surveyors. Singapore's maritime activities are essential in a global economy where 90% of global trade is moved by sea and a significant portion through the Singapore Strait. In 1993, the MPA (Maritime Port Authority of Singapore) designated specific geographical coordinates within their territorial seas to facilitate the exchange of these maritime services. These 'anchorage zones' supported the rationalization of Singapore's maritime economy while making regional and global flows of trade more efficient. Shipowners and managers anchor vessels in these zones for ship registration and marine surveys as well as arranging for the supply of crews, bunkers, food, and fresh water. According to Wilhelmsen (a global maritime industry group), laying up ships costs shipowners up to $1000 a day within "port limits"—a **hefty** sum for shipowners searching to counteract falling rates of profit. Laying up ships in OPL thus became a viable strategy to delay devaluation during challenging times, as seen in the most recent economic crisis.

6.   Driven by "optimism, **euphoria**, and irrational exuberance", the coercive laws of competition and the "China commodity boom years", shipowners commenced a period of largescale ordering between 2004 and 2008. A total of 500 million DWT was ordered, while 1. 18 billion DWT was already in circulation. As in the early 1970s, 1980s, and the mid-1990s, banks extended "cheap money" to shipowners interested in fleet expansion and modernization. By 2004, a **cornucopia** of sophisticated financing options was available through high-yield bond markets, public equity shipping markets, and capital raised in the "speculative-grade bond sector". During this four-year period, shipping debt added $230 billion worth of mortgages, much of which was unsecured and covered 100% on the hull. After the global financial crisis, however, the shipping industry was in trouble. In light of the Lehman Brothers-collapse, it became clear that shipping was facing a radically different future: extensive economic losses, devaluation of ships, and the restructuring of financing options available for the ailing shipping sector. The *2009 Review of Maritime Transport* suggested shipowners nee-

ded to find ways to "manage their losses" and listed as primary options temporarily laying up ships and permanently removing ships through scrapping. In this period, shipowners began to rely more heavily on Singapore OPL as a place to anchor temporarily unemployed ships.

7. The MPA Singapore recognizes two narrow areas along the Singapore Strait as "east and west **anchorages** zones", but makes no claims to govern these spaces. To captains, seamen, marine surveyors, brokers, and traders, these zones are known as "OPL". OPLs are used by shipowners and managers who require maritime services, but do not need to enter Singapore's port limits. Shipowners can anchor there and avoid port and piloting fees. In 2009, it became evident that ships were anchored in OPL for extended periods of time, avoiding daily lay-up fees and delaying economic losses while waiting out the global economic crisis. These OPL-areas were remade into a liminal zone. A place where not yet de- or re-valued ships could be suspended while their owner found new ways to navigate the transformed shipping economy. In 2009, it was estimated that hundreds of ships were anchored in these spaces; by 2014 marine surveyors confirmed that the number is closer to a thousand ships laid-up in Singapore OPL, including new OPL-areas in the South China Sea. Many of the ships were laid-up for months between 2009 and 2014, with only two crewmembers on board. It is no wonder they earned the moniker, "ghost ships of the recession".

(1,482 words)

➢ *New Words*

scrap［skræp］   *n.* things that are not wanted or cannot be used for their original purpose, but which have some value for the material they are made of 废料;废品

They collect and sell scrap metal to cover their rents and care for three grandchildren.

他们收集并出售废金属,来支付房租并照顾三个孙子。

fictitious［fɪkˈtɪʃəs］   *a.* formed or conceived by the imagination 虚构的;假想的;编造的

The characters in this story are all fictitious.

这个故事里的人物都是假想的。

turnover［ˈtɜːnəʊvə(r)］   *n.* the volume measured in dollars 营业额;资金

The store greatly reduced the prices to make a quick turnover.

这家商店实行大减价以迅速周转资金。

flip-flop［ˈflɪp flɒp］   *n.* open shoes which are held on your feet by a strap that goes between your toes 人字拖

The lack of support of the flip-flop also causes pain in the tendons on the inside of the foot and lower leg. It can also lead to shin splints.

由于穿人字拖缺乏对脚部的支撑,也会造成脚内部及腿的下部肌腱疼痛,还可能造成小腿疼痛。

haphazardly［hæpˈhæzədlɪ］   *ad.* in a random manner; without care; in a slapdash manner 偶然地,随意地;杂乱地

Scientists already know that birds don't sing haphazardly, but in a way that is governed by a set of linguistic rules that form strings of sounds.

科学家早就发现,鸟类的歌唱并不是杂乱无章的,而是一连串有规则的语言符号。

stack［stæk］   *v.* to fill something with piles of things 使成叠(或成摞、成堆)地放在……;使码放在……

He ordered them to stack up pillows behind his back.

他命令他们在他背后堆放一些枕头。

asbestos［æsˈbestɒs］   *n.* a fibrous amphibole; used for making fireproof articles; inhaling fibers can cause asbestosis or lung cancer 石棉

All forms of asbestos are carcinogenic to humans, and may cause mesothelioma and cancer of the lung, larynx and ovary.

所有形式的石棉对人体均有致癌性,可导致间皮瘤、肺癌、喉癌和卵巢癌。

residue [ˈrezɪdjuː] *n.* matter that remains after something has been removed 残渣;剩余;滤渣

This releases a mixture of hydrogen and carbon monoxide, known as syngas, and leaves a residue of amorphous elemental carbon called carbon black.

这一过程产生氢气和一氧化碳的混合物,也叫合成气,剩余的是无定形的基本碳,也叫碳黑。

linger [ˈlɪŋgə(r)] *v.* to continue to exist for longer than expected 继续存留;缓慢消失

When apartheid is over the maladies will linger on.

种族隔离废止后,这种社会弊病还会继续。

debilitate [dɪˈbɪlɪteɪt] *v.* to make weak 使衰弱;使虚弱

Schistosomiasis is one of zoonoses in the world, and severely debilitate people and animal's health.

血吸虫病是一种人畜共患病,严重威胁着人类和动物的健康。

tonnage [ˈtʌnɪdʒ] *n.* a tax imposed on ships that enter the US; based on the tonnage of the ship 吨位,载重量;船舶总吨数

The ACP forecasts that, thanks to the expansion, total tonnage will rise from 280 m tons in 2005 (its base year) to 510 m tons in 2025.

巴拿马运河管理局预测,得益于扩建,运河全部通航吨位数将从 2005 年(其基准年)的 2.8 亿吨上升到 2025 年的 5.1 亿吨。

woe [wəʊ] *n.* misery resulting from affliction 困难,灾难;痛苦,悲伤

Such luxury at a time of economic woe may be surprising.

在经济困难时期,如此奢侈使人吃惊。

mortgage [ˈmɔːgɪdʒ] *n.* a conditional conveyance of property as security for the repayment of a loan 抵押;抵押贷款额

This shows you how to create a basic data table that will display a list of all mortgage rates.

这将向您展示如何创建显示所有抵押比率清单的基本数据表格。

exacerbate [ɪgˈzæsəbeɪt] *v.* to make worse; to exasperate or irritate 使加剧;使恶化;激怒

The structure of the fund management industry and stock market may also exacerbate the problem.

基金管理行业和股票市场的结构也可能加剧这一问题。

equity [ˈekwətɪ] *n.* common stock; the net value of the mortgaged assets 普通股;抵押资产的净值

Do you plan to go for debt or equity financing?

你是打算进行债务融资还是股权融资?

mangrove [ˈmæŋgrəʊv] *n.* a tropical tree or shrub bearing fruit that germinates while still on the tree and having numerous prop roots that eventually form an impenetrable mass and are important in land building 红树林

Mangrove trees provide a naturally brackish habitat for fish, but much of Aceh's mangroves have been cut down over the years, precisely to build fishponds.

红树林的树木为鱼类提供了自然的咸淡水栖息地,但是这些年来在亚齐省(印度尼西亚最西部的一个省)却恰恰为了修建鱼塘而将红树林砍掉。

ammonia[ə'məʊniə]   *n.* a pungent gas compounded of nitrogen and hydrogen; a water solution of ammonia 氨;氨水

Hydrogen is used extensively in industry for the production of ammonia.

氢气大量地在工业中被用于氨的生产。

caustic['kɔːstɪk]   *a.* harsh or corrosive in tone 腐蚀性的

In the Draize test, caustic substances are placed in the eyes of conscious rabbits to evaluate damage to sensitive eye tissues.

在眼睛的刺激实验中,腐蚀性物质被注入意识清醒的兔子眼睛里以测试对敏感性眼组织的伤害。

chromium['krəʊmiəm]   *n.* a hard brittle multivalent metallic element; resistant to corrosion and tarnishing 铬(24 号元素,符号 Cr)

And like coal ash produced today, the particles would have been loaded with toxic metals such as chromium and arsenic.

就像今天的煤灰一样,这些颗粒中可能也含有一些有毒金属,如铬和砷。

bunker['bʌŋkə(r)]   *n.* a large container for storing fuel 燃料舱

A bunker is a container for coal or other fuel.

燃料舱是装煤或其他燃料的容器。

hefty['heftɪ]   *a.* of considerable weight and size 异常大的

Police strictly enforce these speed limits with hefty fines.

警方以巨额罚款来执行这些限速措施。

euphoria[juː'fɔːriə]   *n.* a feeling of great (usually exaggerated) elation(常指较短时间的)极度兴奋,情绪高涨,狂喜

Thomson adds that the American market can flip-flop between euphoria and despair in a heartbeat.

汤姆森补充道,美国市场可以瞬间实现狂喜与绝望的跳跃转换。

cornucopia[ˌkɔːnjuˈkəʊpiə] *n.* something that is or contains a large supply of good things 丰盛;丰富;丰饶

From the low-growing Indian rhubarb to the mighty English maples, the plants in this region get ready for the winter with a cornucopia of colors.

从低速度生长的印度大黄到巨大的英国枫树,这一地区的植物用丰富的色彩为冬天的到来做好了准备。

anchorage[ˈæŋkərɪdʒ]　　*n.* the condition of being secured to a base; a fee for anchoring 锚地;下锚

The ship remained in anchorage for a month.

船在锚地停了一个月。

➤ *Phrases and Expressions*

in part 部分地;在某种程度上

in advance of 超过;在……前面

an influx of ……的大量涌入;……的大量汇集

mop up 擦;用拖把拖洗;结束

in trouble 在监禁中;处于不幸中;处于困难中

in light of 根据;鉴于;从……的观点

➤ *Terminology*

fictional capital 虚构资本

slow violence 缓慢的暴力

asset play 资产隙

equity-financing funds 股权融资资金

➤ *Proper Names*

Bangladesh 孟加拉国(亚洲国家)

polychlorinated biphenyls (PCBs) 多氯联苯

polycyclic-aromatic hydrocarbons (PAHs) 多环芳烃

NGO (Non-Governmental Organization) 民间组织;非政府组织

OPEC (Organization of Petroleum Exporting Countries) 石油输出国组织

DWT (deadweight ton) 载重公吨位

OPL (outer port limits) 外港限制

MPA (Maritime Port Authority of Singapore) 新加坡港口管理局

the Singapore Strait 新加坡海峡

➤ *Translation*

1. Gas and oil residues linger in enclosed areas on beached ships... cuttermen light their torches. (Para. 2)

天然气和石油残留物残留在搁浅船只的封闭区域,当切割工人点燃火把时,经常会引起爆炸。

(这里主要是指《环境与规划 A:经济与空间》杂志公布的海洋调查研究结果。)

2. The last reported fatality at the yard depicted was in September... fell, struck, and killed a man working below. (Para. 2)

最后一次报道的死亡事件发生在 2016 年 9 月,当时一块铁板掉了下来,砸中了一名在下面工作的工人,导致其死亡。

3. As devalued tonnage continued to plague the shipping... scrap markets became critical. (Para. 3)

随着贬值的吨位继续困扰航运部门,进入废船市场变得至关重要。

4. Hossain highlighted the hazardous environmental and labor conditions caused by toxic materials released... and moving. (Para. 4)

侯赛因强调了在拆船过程中由于倾倒、溢出、混合、储存、燃烧、火炬切割和移动释放的有毒物质,所造成的危险环境和劳动条件。

5. But with the ever-increasing build-up of surplus ships... to help mop up devalued vessels. (Para. 4)

但随着过剩船舶的不断增加,受竞争驱动的船东们也需要其他场所和做法来帮助清理贬值的船舶。

6. Singapore is a one-stop service center for maritime... and water suppliers as well as ship repair and marine surveyors. (Para. 5)

新加坡是一个一站式的海事服务中心:经纪和贸易公司、船舶融资和保险服务、燃料舱、船员和水供应商以及船舶修理和海事测量师。

7. Laying up ships in OPL thus became a viable strategy... in the most recent economic crisis. (Para. 5)

因此,在外港限制区域建造船舶成为一种可行的策略,可以在困难时期推迟货币贬值,就像最近的经济危机一样。

8. By 2004, a cornucopia of sophisticated financing options was available through high-yield... sector". (Para. 6)

到 2004 年,通过高收益债券市场、公共股票运输市场和"投机级债券部门"筹集的资金,出现了大量复杂的融资选择。

【这个句子结构简单,在翻译时采用改变原文词语的前后次序的换序译法比较恰当,将主语部分"a cornucopia... options"放在句末翻译,使译文在最大限度上自然通顺。当然也可以采用直译的翻译方法。】

9. During this four-year period, shipping debt added ＄230 billion... the shipping industry was in trouble. (Para. 6)

在这四年期间,航运债务增加了 2 300 亿美元的抵押贷款,其中大部分是无担保的,并覆盖了 100％的船只。然而,在全球金融危机之后,航运业陷入了困境。

10. Many of the ships were laid-up for months... only two crewmembers on board. (Para. 7)

许多船只在 2009 年至 2014 年之间闲置了数月,船上只有两名船员。难怪他们被称为"衰退的幽灵船"。

【这个句子是带有介词短语做伴随状语的简单句。在翻译时,把原文中的被动语态转换成译文的主动语态,用转态译法进行翻译,译为"闲置了"使译文通顺自然。】

 **Exercises**

1. Fill in the blanks with the proper given words, and then translate the sentences into Chinese.

recession	devalue	accumulate	scrap
shipowner	rationalization	exuberance	navigate
surveyor	exacerbate	fictitious	expansion
spill	outweigh	equity	designate

1) Greece stands out for the size of its debt stock, the scale of its budget deficit and the grimness of its growth prospects given high domestic costs and an inability to _____.

2) Greek Shipping Minister Kostis Moussouroulis says in Shanghai that Greek _____ has recently signed contracts to buy 142 new ships from Chinese shipbuilding companies.

3) To me, this type of Thai film has some of the _____ and energy of the John Hughes teenage romantic comedies of the 1990s, such as *Pretty in Pink* or *Sixteen Candles*.

4) As a youth, Washington worked as a _____ and in 1754 was sent with a military expedition to maintain Virginia's claim to Ohio lands against the French.

5) Figure 1 and figure 2 below illustrate the varying hourly workload for three _____ applications, and the resulting redundant capacity when they are deployed separately or together.

6) Some experts argue that the broad use of the scanners raises the same question that pertains to any other routine exposure to small doses of radiation: do the benefits _____ the risks?

7) The Code does not specify the type of organization that should serve as the national authority and each Member State may _____ the organization it believes can best fulfil the role.

8) Dark energy makes up three-quarters of our universe but is totally invisible. We only know it exists because of its effect on the _____ of the Universe.

9) All of these are based around one simple fact, to pay off debt, or to save money, or to _____ wealth, you must spend less than you earn.

10) The actions include color changes and text displays based on the values, as well as hyperlinks from specific areas of the diagram so that users can _____ to other diagrams.

11) They should promote the reform of the international financial system, enhance the diversification and _____ of the international monetary system and jointly safeguard financial stability.

12) In reality, gaps in health outcomes will be reduced, and health systems will strive for fairness only when _____ is an explicit policy objective, also in sectors well beyond health.

13) Following the explosion and fire on the rig, the blow-out preventer on the sea bed should have automatically sealed the well, but it failed, resulting in the massive oil _____.

14) The fall in sterling will also blunt deflationary pressures that might otherwise _____ the credit crisis, since when prices start falling the burden of debt rises in real terms.

15) Asset bubbles are as old as capitalism, and since this is a movie about capitalism and the current Great _____, it would have been nice to see some of this in the movie.

16) Smoke rose from a crude earthen forge, and women took turns at a goatskin bellows while men and boys pounded _____ metal on small anvils, shaping it into cooking spoons, axes, and other simple wares.

2. Translate the following passage into Chinese.

Scrap data since 1975 shows that both progressive and periodic phases of devaluation have not only produced increasing quantities of devalued fixed capital but have become part of the logic of capital circulation through the shipping sector. This new scale of devaluation reveals how overproduction is, in part, produced by overaccumulated money capital moving into the shipping sector. This orientation of capital could not "be accomplished without a money supply and credit system which creates 'fictional capital' in advance of actual production and consumption". It is evident that each shipping crisis was preceded by an influx of increasing amounts of fictitious capital used to finance new technologies. These innovations sped up capital's turnover time in and through the shipping sector. This led to capital being advanced faster than it could be absorbed. This brings us to the increasing use of shipbreaking and laying up ships as strategies for spatially displacing economic losses and helping bring about new cycles of accumulation—with corresponding implications for labor and the environment.

3. Translate the following passage into English.

然而,这一解释描述了症状,却忽略了在空间和时间上正在发生的其他动态,这些动态强化了这种"不平衡"的状况:价值是如何在航运部门产生的,资本又是如何在航运部门循环的,这对劳动力和环境产生了一系列实质性的影响。航运市场与市场上使用经验数据报告和实地研究调查收集的数据,本文将谈及两个实践场所,孟加拉国拆船码头和铺设了船舶的新加坡的"外港限制"——作为一种从 20 世纪 60 年代中期开始通过航运部门来分析扩大规模的资本流通方式。